A new book about working, living
and dying
in

DETROIT !

ISBN 0-7414-0719-1

Published by:

INFINITY
PUBLISHING.COM

Infinity Publishing.com
519 West Lancaster Avenue
Haverford, PA 19041-1413
Info@buybooksontheweb.com
www.buybooksontheweb.com
Toll-free (877) BUY BOOK
Local Phone (610) 520-2500
Fax (610) 519-0261

Printed in the United States of America

Printed on Recycled Paper

Published July, 2001

To my Friend who they murdered
&

AMY & MAGGIE

ROGER
&
TEAM PENSKE

WHO KNOW NO OTHER
WAY !

DEDICATION

My name is Peaches, 876, Six Hot Pepper 10, 666, Tango in Question or whatever code name you would like to subscribe to me.

This book is dedicated to the Bravehearts, Oaktree Men and Warriors, who saw how evil the Kingdom of Detroit was and who had the strength, integrity, courage and wisdom to bring it down brick by brick. Theirs was an exceedingly difficult task, fraught with tremendous danger for themselves and their loved ones.

They are true Heroes! The Truck people, the good Car people, the Republicans, the good F.B.I., the Environmental Protection Agency, the good Construction people, the World Health Organization, the Emergency Room Physicians, the good Trauma people, the Pizza people, the Retailers, the Grocers, the Music men, the Mafia people, just to name a few, who taught the keepers of the Crypts a new way of doing business in the city of "TROIT".

Their choice was the "Right Choice", because their courage, guts and perseverance saved the lives of the 1,200,000 people who live here on both sides of the border.

Whoever said there were no Heroes left was wrong! They are alive and well and living here!

This is their story.

Prologue:

Where does one start to tell the story of mice, maniacs and monsters and of the Bravehearts, Oaktree men and Warriors who fought the war to bring the evil kingdom down in the city of "Troit". How does one tell the story of Supreme Court justices who are just asses and captains of industry who poison the land, the people, the water and everything they touch in the pursuit of profit. How does one tell the story of health care systems, huge health care systems that commit homicide and genocide and get away with it year after year in the name of "bidness", business. Systems that own their own hospitals, their own doctors, their own health maintenance organizations and their own off-shore, for-profit insurance companies, whose CEO's create a façade of quality and excellence which is all smoke and mirrors, while murdering anyone who tries to erode the short term bottom line. Or city administrators whose henchmen disembowel and dismember for the fun of it, any one who gets in their way and buries the bodies in neat small packages in the local HP crypt. How do you tell the story of Southern Christian Leadership Schmers who sell out their own people, leaving them to die from tainted water, in crumbling burned out buildings and at the hands of their local "po-lice", while they fly off in Lear jets, raping and robbing the poor, the weak, the sick, the old and the vulnerable. How do you tell of disenfranchised Democrats, who spew fetid spray into the airwaves, like whales after eating squid. Noxious clouds of poison gases, to cover their trail of bribery, prostitution and corruption, murdering the population that they so vehemently espouse to protect. The Democratic dream, the Democratic façade, the Democratic farce. The armed farces, Charles Wasson's armed farces, whose job first and foremost is to protect themselves. Don't ask, don't tell or you'll go to hell. Hell Michigan that is, where they paper walls hang her, put her in the dryer and fry her, erase her, drug her, poison her, murder her, in order to keep silent the realities of the crypts. When that doesn't work, they arrange accidents that are not accidents, killing all of those that tried to help, tried to do the right thing because they knew that someday they would be the victims of the same animals. They are masters at turning Tennessee against Texas, union against management, Arabs against Jews, divide, divide, divide and conquer. Protect the evils, the atrocities and the genocide as long as we are making a profit. Close your eyes and blink, pretend that it didn't happen, make the bodies disappear. The bodies of the young, the old, the poor, the weak, the rich, the famous, make all the bodies disappear. Not here in Democratic America, these things don't happen here they say. This is the land of the free and the home of the brave. We don't set up public officials with hookers, our philanthropist's don't pimp for HIV+ offspring or snow the good guys with heroin and hallucinogen laced pork chops, or set up federal judges with Hord wives on safari or buy-out Roman Catholic bishops over golf. We don't blackmail the chairs of Canadian banks and set up the heads of the World Health Organization with wives who burn their marital beds. We don't. We can't. We won't. At least we won't admit to it, not here, not here in democratic America. Land of the free, home of the brave, filled with pus and pestilence. A land where attorney generals are bullied, harassed and intimidated into silence and bought by big "bidness" or are made to step down, capitulate, in order to preserve the democratic honour. Hizonner the mayor, Archie the Zero, whose city has a higher mortality rate than a third world country. Death before dishonour is his cry, kill them all before they dishonour us by telling our sins. Hide the contaminated water issue, the dysfunctional fire department issue, the police murders, hide the agent orange issue, the gulf war poison gas issue, the depleted uranium shell issue, the carbon monoxide massacre issue, the Packard body repository issue. The guns, the guns they are smoking and the bodies are dead, hundreds of bodies, thousands of bodies.

The bodies of the killing fields, the killing Field's brothers that is. Let's kill the babies with contaminated water, let's formaldehyde the children and elderly on dialysis. Let's create accidents that are not accidents. Falls, suicides, homicides, but let's do it scientifically, Pisser scientifically, so that it is almost impossible to detect. Let's murder racecar drivers with a conscience and hang world health care workers from the bridge. Let's poison the bleeding heart liberals with warfarin, coumadin, rat poison to make them really bleed. DIC bleed, from their eyes, their nose, their mouth, their pores and make it look like an accident. Let's cut off their hands and crush their larynx and dismember them with surgical precision, then trash the place so it looks like a hit, a mafia hit, so that no one will know that it would be one of Archie's hits, Benny the Rat's hits, Prick Stein's hits. Let's commit atrocities, the more demonic the better, they will never link it to the city. Never in a thousand, million years link it to the city and a former Supreme Court "justass". Don't ask, don't tell, Archie don't know, Benny don't know, Charles Wasson don't know. They know nothing as long as Archie's war chest is full and they can sit and watch pornographic films of all the women they have drugged in the basement of Hooligan Manor, Archie Bunker's basement, the biggest crypt of them all.

Contents

The BAD GUYS

Contents

The GOOD GUYS

The Bad Guys Revisited!

THE BAD GUYS

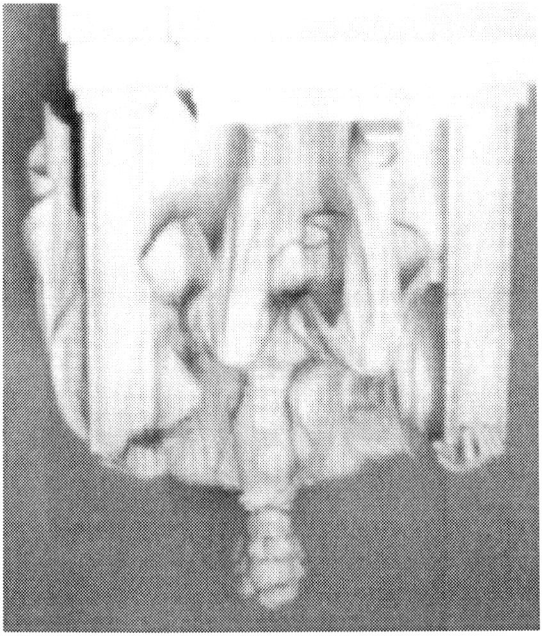

THE KEEPERS of the CRYPTS

Chapter 1
Who Killed Hoffa? – Despeisner's Tale

I know that this is reaching back, but this is the turning point, the beginning and the murder that gave them the sense of power that they could butcher anyone and get away with it, so it is important to include it here. Jimmy did die at Macchus Red Fox, but not in the parking lot. He was murdered in the long, basement dining room that you reach by the narrow winding stairs to the left of the entrance. Fendry and Maxie planned it, because Hoffa was such a pain in the ass, such a f...king pain in the ass, that they needed to get rid of him, had to rid of him, because he was costing them too much money. But they had to make it look like an outside job, a mob job, a teamster job. So they invited him to dinner just to "talk" and the moment he got to that basement door, they ambushed him, dragged him to the table and shackled him to a chair by all fours. Max Pisser was the brains of the outfit, he planned it all with surgical precision. They even brought in a young surgeon from Fendry's own hospital to facilitate the process. A young struggling trauma surgeon named Snobeid with a lot of mouths to feed. He was a scared, penniless Lebanese boy and they promised him a bonus, a huge bonus if he would facilitate "a surgery" and Snobeid obeyed.

Hoffa was a tough son of a bitch, but no match for Fendry and Maxie's hired henchmen. There were a lot of them too, McNutty their 300 pound finance boy was there, the one who financed their new ventures from the proceeds of Fendry's new hospital, Tony the Boss Provenzano and a young Jew named Despeisner who had a reputation for killing anything he touched. They pinned Hoffa in the chair and gagged him with a filthy dinner napkin and directed Snobeid to earn his bonus and Snobeid obeyed. Slowly, methodically, he applied tourniquets to all four limbs so Jimmy wouldn't bleed to death prematurely. They wanted to savour the moment. Jimmy had cut them to ribbons financially, so they were going to cut him to ribbons literally and figuratively speaking. They nodded to Snobeid and he began to methodically amputate first Jimmy's hands, then his arms, then his feet, then his legs. All the while feeding them into the freaking industrial strength meat grinder that they had brought down from the kitchen. Fed his f…king hands, arms and feet into the f…king meat grinder right in front of him and they watched while Despeisner's dogs, one big black and two big brown dogs ate the f…king hamburger as it coiled out of the meat grinder. They wanted to show him that this is what they do to barking dogs, dogs that bark. The dogs strained on their leash with Jimmy's congealed blood hanging in ribbons from their chops, eagerly waiting for more. But Jimmy was a tough mother-fucker, he wouldn't die, couldn't die, didn't die, even when he was left a stump-less torso with four tourniquets on to prevent him from bleeding to death too soon! This enraged the freaks from the crypts, so they threw him on the table and cut off his balls and cock and fed them to him, stuffed them in his mouth, so that he would gag and choke on his own balls and then Jimmy died. Died in a pool of his own blood and they tossed his torso on the floor so the dogs could finish him off, gnaw on what was left of him, to make it easier to feed into the meat grinder, what was left of him and soon all that existed of Jimmy was a stain on the floor, a bloody stain on the floor of Macchus Red Fox. Then they had the contractor come in and pour a new three inch floor of concrete over the stain on the basement floor and by the next afternoon, when Jimmy was noted to be missing, the old carpet had been laid back down and Jimmy had disappeared forever. No one spoke about it, Snobeid was rewarded handsomely, made chief trauma surgeon and given fancy cars and houses. Despeisner had his way paid through law school and whenever he got out of line they said he was an alcoholic and on drugs and threw him in the looney bin. McNutty was made into the Chief Financial Officer and forever after he would sit like the cock of the walk at the head of the same table with his legs spread against the table legs where Jimmy Hoffa had died. I know, because I sat in that dining room with them many times. It gave them a tremendous sense of power to know that they had offed Hoffa and gotten away with it. Now every intimate celebration was held in that same dining room with McNutty presiding like the King at the head of the table knowing that even after his cesspool of a hospital did in Fendry, that he still reigned because he had the keys to the castle, the castle crypts, Hoffa's crypt and in the eyes of big "bidness", that made him a very powerful man. This was the beginning.

Fendry's death was number two.

Chapter 2
Oh Fendry!

Fendry the Duce led a charmed life, scion to Fendry 1, he was born into a world of privilege, power and expectations. For the first half of his life he did the right things, marrying a helmet haired blue blood and siring a litter of little prince and princesses. His life consisted of attending soirees and charity balls and dabbling a little in the family "bidness" on the side. His business skills were no match for his father's however and gradually his power in the company eroded, and it drifted into more rigid and controlling, comp-trolling hands. The Duce became bored and sought out the company of younger women, prettier and more playful women and settled on one with two young ones of her own, dumping the helmet head and alienating his first litter.

Life for the Duce with his new acquisition was smooth sailing until a little medical problem developed, so he checked into one of his own hospitals. There he promptly caught legionella from the air vents that probably had never been cleaned, but certainly would be now. Fendry wasn't as tough as Jimmy though, years of hard drinking and fast living had taken it's toll and his lungs rapidly filled with fluid from the legionella organisms that clustered like grapes in the sacs of his lungs. He couldn't breathe, his heart started failing, his kidneys felt the strain, the Duce was in trouble so they transferred him to the big house. Chez Fendry his own center of medical excellence, where all the medical expertise in the world waited for a chance to cure Fendry and in doing so ingratiate themselves forever to the Hord family. His lungs failed, so they put him on a ventilator, his heart failed, so they put him on potent intravenous drugs, his kidneys failed, so they put him on one of their miraculous new dialysis machines and promptly killed him. Killed him with a bolus of contaminated water. His immune system already weakened by the legionella from the vents at the little house, promptly succumbed to the water borne serratia and pseudomonas bugs at the big house. The endotoxins from the gram-negative bacteria overwhelmed him and Fendry soon became a casualty too. Died at his own hand in a manner of speaking, because in Fendry's hospitals profits were king. They were organized that way. Fendry owned the building, owned the doctor's, the nurses, the HMO, the offshore for-profit malpractice insurance company, the works. It was a closed shop, a closed loop, you didn't need to spend money on cleaning air filters or dialysis water membranes, that would cut into the bottom line. So Fendry, oh Fendry, poor Fendry died in a manner of speaking at his own hand, because he wouldn't spend a few bucks to do the right thing. He chose the short-term bottom line over long-term quality of care and it killed him, promptly killed him! Fendry learned the hard way.

His choice was wrong!!!

Chapter 3
Perversified's Die-Alysis

Perversified is where I first learned of Fendry's demise. I had been hired first as a dialysis nurse, then as an educator, then as the nursing administrator for their programs. When I learned of how they had did in Fendry Hord, I resolved that we would do a better job on our watch. One we could be proud of, that would protect the welfare and well being of our 1600 patients in four states, so that no one else would die like Fendry had. It was a lot of hard work, tough work, particularly because our medical director, the man we all revered left and went to New York, tired of the daily battles of fighting for quality of care against the CFO's whose only concern was today's bottom line. Our new medical director had only a cursory interest in dialysis. He focused instead in selling himself as the Guru of electrolytes and selling the programs for his own gain. This made the job even harder. The die-alysis business was an extremely profitable one for Chez Fendry. A $64,000,000 a year cash cow that kept the hospital and the Hay Hords floating. Our former medical director had been a wizard at combining excellent clinical practice with excellent financial deliverables, but he was trying to push a new technology, a much, much safer technology for patients and staff, one that didn't use the cancer causing toxic agent formaldehyde. It was heat sterilization, we had proved that we could turn a good profit with it, but it wasn't good enough for the Sheepherds and Bevie Hagmans that wanted their 30% bonus that we all got at the end of the year based on profitability. They wanted it to be a <u>big</u> check. They weren't even willing to let us play with the process internally to refine it for use in all of our programs. Oh, they let us sell it all right. We went all around the country peddling heat sterilization to facilities that did care about their patients and staff. But we couldn't use it at home because it didn't make us the $64,000,000 that formaldehyde did. So Lathan Nevine left for New York and I stayed three more years until the hypocrisy got to be too much for me too. When you couldn't buy a $130. blood sugar machine for a unit where sixty percent of the patients were diabetic, even though that unit made $8,000,000 a year. We couldn't buy it because the

McNutties, the Sheepherds and the Hags were worried that such reckless expenditures would cut into the bottom line, their bottom line, their check at the end of the year. So after three more years of watching them replace our professional staff with unskilled care givers and watching patient's treatment times be deliberately cut to cram as many treatments in as possible to optimize profit, despite the fact that it harmed our patients. After three years of watching our patient's being injected with the embalming fluid formaldehyde each treatment, even though we knew that our death rate was 25 to 50 percent higher than Canada, France and Germany, who didn't re-use artificial kidneys, I left too. In December of 1994 I left and things have never been the same since!

Chapter 4
Mercy, Mercy, McNutty

My first experience with the wrath of Pom McNutty was at Mercy Hospital. I had been told that Pom was a psychotic nut who threw furniture and people around rooms and that he would get so enraged that his false teeth would literally fly out of his mouth and land on your desk. That wouldn't stop him though, he would just cover his teeth with his big paw and continue to spit all over you literally and figuratively speaking. Pom was huge, built like a refrigerator and the ground actually shook when his three hundred plus pounds came fe, fi, foing into a building. Everybody was afraid of him, terrified of him because they knew of his involvement in Hoffa's death. They knew how Jimmy died, where he died and who had orchestrated it and they didn't want the same thing to happen to them. Pom cared nothing about patient care delivery, only the bottom line and he would have anyone killed who stepped out of line and disrupted that bottom line. I had first met Pom at one of our quality management seminars. Images, smoke and mirror images were very, very important at Hord. So we made a big hoopla about espousing the principles of quality management, weren't allowed to use it of course, but we certainly made a lot of noise about it. The funny thing was though, all the things they taught us about Deming and Drucker and Ohmae made so much sense. Particularly in health care where failures were measured in human lives and quality of life, not just dollars and cents. How much better, smarter, more humane to have care delivered by skilled people, people who knew the principles of asepsis, than to cost someone an arm or a leg from a hospital acquired infection. Each hospital acquired infection added on average, $60,000 to a patient's hospital stay according to C.D.C. Sixty thousand could buy a hell of a lot of skilled care givers. But Pom didn't see it that way. He felt nurses were a pain in the ass, bleeding heart liberals who cost him money. The only reason why he kept us around was number one he wasn't smart enough to do our job and number two we did all the work. The only people he despised more, were the Drants. That's what he called them, Dr. ants who worked their tails off at Chez Fendry to make Big Pom and the Hay Hords wealthy men. After I left Hord's, one of my first jobs as an independent consultant was with Mercy Hospital. Despite the fact that Hord managed the Mercy dialysis programs out of Ohio, the programs were in big trouble, in danger of being decertified by the state. So Mercy asked me to come and help them and I did. We had fun, so much fun there. The patients and the staff were both tremendously disadvantaged economically, but they had such spirit! They knew that Hord was trying to shut them down and take over the program to shift the revenue over to McNutty and Chez Hord. It was very difficult, we had no financial resources to work with, but using the principles of quality management that we had been taught at Hord (but never allowed to use), we wrote their policies, educated their staff and got our pharmacy reps to sponsor little prizes for our patients. Little things to reward them for helping take better care of themselves. Mittens and socks and hairbrushes. Food for those that had none, a jacket for the peritoneal dialysis patient whose supplies we had to leave in the bushes because he was homeless. For the staff, we charted their progress in achieving excellent outcomes on the wall for all to see. We fired the security guards who were selling drugs to our patients in the lobby and got rid of the ward clerk who was stealing the reimbursement money that was supposed to go to patient care. With that money, we painted the place and paid for little lunches to thank both patients and staff for all their hard work. We made a raid on Hord's Kingswood school warehouse and shamed Fendry into giving us some decent furniture for the waiting room and patient care areas and we turned the program around. In less than six months we turned the program around clinically, financially and in spirit. This pissed off McNutty so much that he hopped in his hearse. (We called his big brown station wagon the hearse, because it killed everybody that got in its road). The very afternoon that the state gave us back our full certification, he hopped in his hearse, convened a full meeting of the Mercy board and had me fired because I had used what I had learned at his house to help someone else. It pissed him off so much, delayed his acquisition of Mercy so long, that he put his 300 pounds in his hearse and promptly had me canned. That was the beginning of my dealings with Pom and I have had his 300 pound shadow following me ever since.

I do strongly believe there is a God however and that he does work in mysterious ways. I spent the summer sailing and reading and by the end of the summer, the Joint Commission for Hospital

Accreditation was back at Mercy and Hord of course wasn't helping them because they wanted to see them closed to shift the business and the revenue down town to shore up Fendry's crypt. So Mercy called me back again and we cleaned up their Emergency Services, Intensive Care and Cardiac Care Services and we won again. Our programs received commendations by the Joint Commission on Hospital Accreditation. Then once again McNutty got in his hearse, flew out to his hearse (it's funny to see a 300 pound bull run) in a maniacal rage, spit his teeth out on the table at the Mercy board and had me fired again.

Chapter 5
The Petroit Murder Center

From Mercy, I went to work as a nursing supervisor for the PMC. I was smarter this time, I wore a ring and pretended I was married. I knew that the chance of me having a Hord sponsored accident that was not an accident, was much less if they thought I was not alone. We had fun there too. I worked on my graduate degree during the day and ran the halls of the hospital at night, trying to bring order into the chaotic gun and knife club environment there. It was a very factionalized hospital, different nursing groups populated all four ICU's. Some loyal to the old Samaritan hospital, some to the old Mercy, some to the new Grace and the militant blacks only to themselves. It was strange to see such divisions, particularly when the patients that we served were all essentially the same. You would have thought that the afro-American staff would have if anything more compassion for our predominantly black patients, but it didn't matter, all they cared about was doing the minimum amount of work for the maximum amount of pay. Thank goodness that philosophy didn't persist throughout the building. The staff in the Emergency were wonderful despite the fact that every night was like one flew over the cuckoos nest. They handled the stabbings, the shootings, the beatings beautifully. They worked as a team constantly re-aligning their finite resources to make the best of every bad news situation. I learned so much there about how to re-organize work flow in an ER to make every precious second count and how to resuscitate people who had literally bled to death from their wounds. I probably would still have been doing this job, if the fates hadn't snapped me back into the complex and frightening world of American "die-alysis".

PMC's dialysis programs were being managed by a guy named Harry Groandin, who by the way had also worked for Hord at one point in his life and had fled to Chicago to stay alive and reasonably well. Chicago had tossed him out, so now he was back in "Troit", doing to the PMC what Hord did to everyone else and PMC wasn't liking it. Their programs were in danger of being decertified too, so they asked me if I would help. What a mistake that turned out to be!

The first couple of years were fun, despite Harry's rage we turned their programs around and grew them from one to eight programs in two years generating $8,000,000 in net revenue for the PMC through the efforts of a little team of only 34 people. If you do the math that's revenue generation of almost $235,000 per employee, per annum, not too shabby for a program that had been haemorrhaging red ink only two years earlier. Most importantly, the patients thrived, the staff thrived and so did our partners. So well in fact that we were able to steal away two programs right from under Hord's nose. This coupled with the fact that PMC purchased Hinai Hospital spelled the beginning of the end and helped me understand why they called Detroit the Brazil of the North and the murder capital of the world. But it wasn't the street crooks you had to worry about or the Mafia people (the Italian mafia). At least they had a code of ethics, they didn't murder women and children. But big business in Detroit did, murdered anyone who got in their way with the help of the animals from Archie's bunker.

They prided themselves in getting away with their atrocities, they always had. But this time they murdered the wrong person, they murdered my friend who worked for the federal government and their lives have never been the same since either!

Chapter 6
The Hinai Holocaust

PMC purchased Hinai with two goals in mind, number one to close them down because they were stealing "bidness" from their hospital down the street and number two, to rape Hinai for the $94,000,000 Hinai foundation money they had in their coffers. Hinai was a freaking mess clinically, physically and financially. Even Holumbia AOL, a corrupt American corporation that bought anything and had made a sweetheart offer to buy Hinai was bailing out. This left their CEO a piano player (lawyer), named Honberg easy pickings. PMC waved a new car, a country club membership and a few bucks under his nose and bought him out like the three-dollar hooker that he was. Honberg knew that part of the deal was the fact that they would close Hinai in a year, putting all the staff out of a job, but didn't care he already had his thirty pieces of silver, so he took the money and ran.

When PMC purchased Hinai, one of the things they asked us to tackle was their concern about Hinai's high death rate. The infection rate in their dialysis program was four times what it should have been and the death rate was twice the national standard. We needed to know why as quickly as possible and so we looked and what a stinking mess we found. There were multiple, multiple issues dating back several years. First and foremost was the incoming tap water from the Detroit water and sewer lines that was so f…king filthy you could almost walk on it. The under- ground pipes were so old that the sewer water and clean water supply were leaking and mixing together underground making the water truly shitty, literally and figuratively speaking. To make matters worse the surrounding industries had dumped so many chemical contaminants into the water that the mercury and aluminum levels were 150 times the safe limit. No wonder half the people in the area were so crazy, they were being poisoned on a daily basis by the water. The city of Detroit had known this for years but the employees would much rather sleep in their trucks all day and Archie would much rather spend the city's money on services for his own personal projects, than clean up the seething cesspools that lurked in the old neighbourhoods. The second disaster that Hinai owned was the fact that there were no back flow preventers on the taps so that every time the staff dumped hepatitis or HIV positive body drainages down the sink, it was siphoned right back up and dumped out the tap on the hands of the next person who tried to wash them. No wonder the patients were dying like flies, the whole place was a sewer. That was another problem very few windows at Hinai had screens on them. The board didn't want to spend the money to replace them, but in some ways the patients who did have flies come in their room were lucky. The flies laid their eggs in the patients infected, rotting flesh and then the maggots kept the wounds clean. Some patients even survived and went home, because no one changed their dressings and the maggots thrived on the rotted flesh, cleaned out the wounds and the patients survived.

The last very chilling issue at Hinai did involve their chiller, a water bath that surrounded a radioactive core in their nuclear medicine department. The shielding around the radioactive core had leaked for years as well, making the basement of Hinai a freaking Chernobyl. Rendering the staff that worked their sterile and causing them to die of cancer by the dozens or die of bizarre infections from the super bacteria that had mutated within the sewer/water pipes in the nearby walls. Hinai was the Jewish community's jewel alright, but it was a black diamond, a cursed pile of stone, that sickened and killed almost every one who entered it!

Despite the severity of these issues, they in and of themselves shouldn't necessarily have been a factor in the death of the dialysis patients. The rules for water treatment were very strict and the special water purification machines should have been able to strain out even solid waste. But there was a huge problem there too. Hord who had been doing dialysis under contract for Hinai, who taught everybody else the standards for water treatment, hadn't been doing a freaking thing.

Perhaps it is best first to explain that in dialysis, water has to be treated as a medicine, carefully purified of bacteria and high levels of chemical contaminants that could so easily kill the patients because it intermixes with their blood through a tissue thin screen or membrane during a dialysis treatment. Hord knew this well, so very well, they were experts in the field of water treatment, they knew what had killed Fendry, and they had done nothing!!

In Aug/97 when we started looking at why all these people had died, why they were still dying, we found that unsafe endotoxin levels from Dec/96 had never been corrected. In addition, the bacterial counts that should have been done at least monthly had only been done four times in the last nine months. The machines that showed contamination and it had to have been gross contamination because Hinai was using the wrong medium to grow the cultures in. The machines that had showed bacteriological contamination had never been cleaned, never corrected and so each and every patient who received the first treatment in the morning received a huge bolus of contaminated water. Shit stew so to speak, sewer sludge, water from the black lagoon and they were the ones that were dying like flies.

It didn't stop there however, there was another factor that became a very important factor when the cover up began, a very, very important factor. The water analyses for chemical contaminants were very disturbing, grossly disturbing. Showing aluminum contamination 150 times the safe limit, as well as mercury and many other toxic chemicals that caused cancer, brain damage, birth defects, heart and nerve disease amongst many other things. The industries surrounding Detroit, primarily Hord's industries were dumping all of their chemical wastes in the clean water supply. Mercury from their ignition switches and aluminum from their engine plants, truly giving new meaning to the phrase, "Think Hord First", especially at autopsy!

So we went to the doctor's at Hinai, the ones who had been directly responsible for providing oversight, to make sure Hord had been doing its job. We showed them all the evidence as to why the patients had died, were dying and they sat there mute like the Drants, Despeisner said they were, afraid to address the issue because they might get sued! So we tried another approach, we got their infection control people together with PMC's legal counsel, who had just dealt with three baby deaths from contaminated water at the other hospital down the street. We gave them all the regulation from the seven different US agencies that said you must immediately correct and report such issues and the answer came down that we would, "Do nothing to trigger an investigation". So we tried another tactic, we arranged a meeting of all the players PMC, Hord and Hinai. The kidney doctors, the infection control doctors, the works. We gave them all packages showing report after report of contaminated water and what we needed to do to fix it. Hord sent their corporate whore David Sheepherd to the meeting, who true to form presented slide after slide of falsified negative results. The Drants looked down at their reports knowing they were as complicit as Hord in these deaths because they had not done their jobs either. The Drants looked down at their reports, sat on their hands and said nothing.

Now believe me, this is not a little cookies issue, a death from negligence is just that, a negligent death. But when you cover up a negligent death, it becomes at best manslaughter or at worst homicide and trust me the thought of spending the rest of my life in Jackson prison with the people sitting around the table was not an attractive one, so we decided on a plan D. This time we sent letters to all three Chief Executive Officers and the Boards of Trustees of all three health care systems. What a farce that turned out to be! To put it gently they were not amused and very clearly not interested in doing the right thing. They convened a tribunal meeting the next week, hauled me downtown in front of a bunch of their corporate flunkies, including the CEO of PMC's girlfriend and proceeded to tell me where I could and could not go and who I could and could not see, then repeated once again that, "We would do nothing to trigger an investigation". Now at that time, I was a member of the board of trustees of the state kidney foundation and president of the state nephrology nurse council and I didn't take kindly to being told where I could and could not go. Lincoln had freed the slaves over a hundred years ago and I wasn't about to become one for a pack of corporate whores! So I explained to them one more time, slowly, softly, quietly, what we knew were our moral, legal and ethical responsibilities in this circumstance. I thanked them all for their time and told them that since we had clearly exhausted all avenues internally to address this issue, that we now had no other choice but to seek external legal counsel. I picked up my little briefcase that had my mini-cassette recorder playing quietly inside, walked to the door and left them sitting in stunned silence on their big fat asses.

I knew what was coming next, I had watched them do it so many times before to many other people. So as soon as I got back to my office, I explained what was happening to our staff and we loaded up all of the proof in the trunk of my car, including all of Hord's and Hinai's bills where they had defrauded Medicare and Medicaid. We loaded it all up, I called an attorney and we promptly took it down to the Federal Department of Health and

Human Services and told them the whole schmer. That was such a funny experience too! Number one, they actually had secret service agents there, people who used electronic voice altering machines and phoney names and the second very interesting but frightening dynamic was that they were only concerned about the Medicare and Medicaid fraud. I can remember sitting in shocked silence across from them after explaining again and again, that we are going to lose more patients if we didn't do something soon. They gave me their cards, arranged for me to meet with the F.B.I and warned me to call them if I started experiencing "problems". I got up thinking now what could we do, hopped back in my car and went to plan E.

The wonderful thing about the international dialysis community is that there are a lot of people out there with ethics, who did care. My experiences with John Hopkins, the Center for Communicable Diseases and Beth Israel, New York had taught me that. So I fired off a letter to every dialysis regulatory group I could think of giving them the details, so at least if I died the next week from an accident that was not an accident, someone might come and make them fix the issues. It didn't take long for the evil doers to get wind of what I was doing, they have an internal process where they immediately start tapping phones of anyone who they perceive as dissidents. Less than a week after my tribunal meeting, two of the head office corporate whores showed up at my office and said they wanted me to take a little vacation while they sorted things out (buried the bodies that is). They asked for my keys and pager, which tickled me because I had never gone on vacation before with such a request being made. I knew that it was what our Philippino nurses called being sent on "Fuckation". I said goodbye to our staff and patients and left.

The next day was when the fun of the exercise began in earnest that was the day of my appointment with the F.B.I, who were also on the 33rd floor of the same federal building. That turned out to be a real hoot in retrospect. All the watchers who were supposed to be monitoring the other watchers were in the same building, making it very convenient for them all to spy on each other. Then to top it all off the Shameritech building, whose president sat on the board of trustees of PMC was right across the street. Shameritech, like Ed and the two temps at AT&T who made its billions off big business were dutifully snooping on all the federal and state agencies to make sure that no one got out of line and disrupted their bottom line. The other stinker in the pile was Homerica, who kept track of all the federal and state agents comings and goings by photographing them at their cash machines.

Clinton, Gore and Reno kept the federal agents on a short leash, limiting their cash withdrawals to one hundred dollars at a time, so that they knew where they were and what they were doing. After the Lewinsky fiasco, they did everything they could to make the government officers tow the line, including exposing and betraying 900 F.B.I. Officers in Clinton's last few days in office.

That was the main reason why Homerica was so successful, not because they were the best bank, but because they kept the democrats informed where the feds were and what they were doing. That way, they could outwit the feds and redirect their energies to other pursuits, like raping Bill Gates and Al Raubman for soft money for their campaign war chests. Leaving them, the Hords, the Pissers, the Ewegenes, the Bunkers, the Crackers and the Jesse Jackass Coalition alone. Knowing all of this, I had begun to think that I was stuck in a B movie, later I knew that the realities of the situation would nauseate even Steven King.

The visit with the F.B.I. was kind of fun, first of all they had baseball trophies all over the place and I can remember thinking that it must be kind of hard to work undercover and have your own softball league. It seemed like an oxymoron, how can your identity be a secret if your team plays league games every weekend of the summer. However, I assumed that they must have known what they were doing, after all they were the F.B.I. So I explained everything to Elena Wolzniak and company about the contaminated water issues, about the fraud issues and about the fire safety issues in the units. They too were most concerned about the dollar issues but at least they did express some genuine concern about the patients and promised they would bring the state in right away. I went away from that meeting feeling slightly better. However, on the way home that night, my 300 pound shadow was back in the guise of a little beat up orange Hord escort that has been following me ever since. Pom McNutty pooped his pants when his spies saw me walk into the 477 building and get into the elevator to the 33rd floor. So they quickly started to plan, Hord had several people on the state licensing board, who were in their pockets for one reason or another. They knew that if they threw a cheap car lease at them and a couple of bucks, they could buy almost anything. So the state did go in, in the guise of Flora Burquart, who used to work for Hinai, who worked with Schmancy Boobola, a Hord " die-alysis" whore who had also worked at Hinai and of course found nothing wrong, saw nothing wrong, because she had been paid like the three blind mice, to see nothing, hear nothing, say nothing. Hord trumpeted this to the press there is no problem! Knowing all the while, that their freaking negligence had cost countless lives.

Once again however, God intervened in a mysterious way. Snowman Hung had been admitted to Hinai hospital with a cold, a minor respiratory problem, a little pneumonia. But once he had been admitted to the fetid cesspool at Hinai, he rapidly deteriorated and had to be placed on a ventilator. They placed him on a vent, filled the cascade, the warming and humidifying unit of the vent with tap water and promptly killed him too! Killed him with a bolus of filthy water to his lungs, which no combination of antibiotics could eradicate. Now they were really pooping their pants. Snowman was an icon in the black community. Their hero, their mentor, their guide. What the hell were they going to do now and while they sat and stewed, we planned another tactic too. The families of the dead patients were shocked and appalled by the state's stance, they had seen for themselves, their loved ones get sick every time they went on "die-alysis". So they started bringing me the charts to review and we started to plan to take PMC, Hinai and Hord to court, to force them to correct the issues. I knew almost all of our hundreds of patients personally and I was going to be damned if we let these assholes massacre them too!

Our case was brought forward by a young ex-marine who a) unfortunately was broke and b) had fertility problems, something that proved to be a factor, which Hord leveraged to their favour later on. The first hearing of our case went up before a federal justice, who had the audacity to tell me it was not my job to protect our patients. I couldn't believe what I was hearing from him. He had been around a long time, he knew what we were dealing with and he said it was not our job. When I countered him, asking him whose job was it then, he capitulated, even though Hord, PMC, and Hinai were all pressuring him to throw the case out of court, he capitulated and gave us a hearing date. At that point, I knew that he knew exactly what we were dealing with. Hord's tactics with him got kind of funny after that. They knew they couldn't buy him with just the usual bribes or a three-dollar hooker, so they sent him on safari with Jac the Knife's wife. Jac, who they all hated because he was such an arrogant little bastard. They sent him on safari with Jac's wife, and she being the good little Hord mossad muppet wife that she was, promptly drugged the federal judge with a potent amnesiac, aphrodisiac named Fentanyl and blew him, literally and figuratively speaking, so they would have evidence, pictures to keep him quiet too! The Cardinal they took care of in a different manner, they took him to play golf and sprinkled a little cocaine on his food so he would follow them like Pavlov's dog and do their bidding. Why they even had him bless Hord's Jewish Nursing Home that was being run by one of the Applebum boys. Drugs were no problem for the Hords, they could get the legal stuff from the Applebums and the illegal stuff from Benny and his Jets. As a matter of fact their most successful "operations", were on people they hooked on heroin or whose brains they fried on hallucinogens. Now perhaps you understand why they call them the sansibelt Hords, because there is no limit as to where they hit people and who they will hit to keep the Hay Hords in football money and their Fanny's in flatties!

During this time or actually before this time, two other sentinel events happened. A key patient Marple Fubbs was poisoned with formaldehyde by Hord then killed by the intractable infection that she had acquired at Hinai. On the day they let her die, the day they could do nothing more because the mutated, irradiated, untreatable strain of vancomycin resistant enterococcus bacteria was killing her, another accident that was not an accident occurred. They tried to murder my brother!

The Marple Fubbs case was a good news, bad news case. The good news was that it was the case that helped us get all of the federal agencies to understand the vast complexity and the extent of the corruption in the crypts of Detroit. The bad news was that one of her daughter's sold her mother and family out like the three-dollar hooker that she was.

When we saw that the state was under Hord's thumb, we decided to go a little farther afield to get the message out. It was fun auditing the charts, Hords had deteriorated rapidly since I had left and now each treatment showed four or five, sometimes up to fifteen different errors in care which at best would compromise care and at worse, readily kill someone. There was a tremendous amount of work to be done and rapidly too. So by day, I would audit charts then fax my work to the house of the three dollar whore, who would in turn leave it on her porch, and the beloved ERP's, the ERP's, the WYATT ERP's, the emergency room doctors and the trauma surgeons who worked all hours of the day and night too, would come and pick up my stuff, type it, validate my perceptions, then fax it back to me the next day through the whore. None of us knew at that time she was a whore, who was on the take from Hord, PMC, Hinai and the city of Detroit. What we knew was that Hord, Hinai, PMC and the city of Detroit would murder the good docs too if they had any direct contact with me! Me, who knew too much, so they had to isolate me until they could kill me. That was difficult too, under the circumstances, they had to really make it look like an accident that was not an accident.

So every day I worked on the charts and every night they were faxed to our friends and together we produced tons of documentation, irrefutable documentation. We even had original copies of charts, thanks to the ERP's. We had everything, the original care plans that showed nothing had been done, the original electrocardiograms, the original x-ray reports before the records were taken to Hord's Alter road record workshop and altered. We even had the multiple practice copies where they had practiced signing dead patient's names to consents for treatment that had never been signed and we tried, boy did we try to get the cases to court, to protect everyone from the carnage continuing. But they were all scared shitless. The huge malpractice firms who weren't afraid of taking on the giants one at a time were certainly afraid to take on all three of them, plus the animals in Archie's ark. We even tried an out of state firm, but they weren't going to do anything to damage the black community and let the world know that his cousin the Clod was responsible for Snowman's death too! So we had to be content with sending all the documents to the regulatory agencies and one by one, Detroit tried to snow them too! Hord was scared and pooping it's pants and had sent out a request for RFP's to sell their dialysis programs. But no one would touch them, they stank like the polluted water that cascaded out of the taps and killed their "die-alysis" patients. The dialysis community knew that Hord had tried to blame the deaths on everyone else, anyone else, including the people that made the kidney machines. So every one stood back and watched them lie, one by one, to all the regulatory agencies who came to fix the problem. One by one, they lied to the Joint Commission on Hospital Accreditation, the Center for Communicable Diseases, the Environmental Protection Agency, the Occupational Health and Safety Board, the Surgeon General, they lied to them all, not knowing that all along the regulatory agencies had copies too, of all the originals from the charts and they had been waiting, just waiting for the asshole windbag Snale Horden and his piano playing whores, Honberg, Snellon, Schnellburg and the Bunkers to perjure themselves and they did. So the corporate whores tried another tactic, played their trump card so to speak. The democrats desperately wanted the black vote, needed the black vote in order to continue their charade of governing. So they brought Gore into the pile, Gore who directed the FBI, or so he thought, to create a little leverage, a little top down pressure so to speak. On the day the issue hit the paper, the front page headline of the paper, the Democrats had issued an edict to make Frank Pelley step down. Frank the eternal general was forced to announce his resignation that very same day, to bump the dialysis story and protect big "bidness". Protect the Hords, the Pissers, PMC, Hinai and the Bunkers. To keep the whole world from knowing that they were killing, literally and figuratively killing the 1,200,000 people who lived on both sides of the river with the fetid stew of sewage and toxic chemicals that they dumped into it daily.

This was the Democratic way, at least the Cracker Coalition democrats, who thought nothing of exposing their own troops to cancer causing radiation by using contaminated depleted uranium shells because it was cheaper. Who had them off a bunch of kids with carbon monoxide on Lake Powell because it was disturbing one of their generals. The democrats who planned the murders of young John and Carolyn Kennedy with precision engineering, hydraulic engineering because they were magical and posed a very real threat to Hillary's New York senate seat that had been bought for her in return for favours to Mossad (the Israeli spy organization). The same senate seat that Kennedy had planned to run for, so the Cracker democratic regime with help from the mossad muppet gang and Powell's jamaican ivory gang, offed them too!

No job too big, no job too small, if it threatens the democrats they kill them all. Killed the sons of the race car drivers with Hords help with precision engineering, on the second turn, two months apart, both the same way, by putting a small spot weld on the throttle that would expand and stick when it was hot. Two,

two, two, get it "with love from Hord"! Making the throttle stick so that as they entered the turn they would lose control and hit the wall. Accidents that are not accidents, precision engineering. Undetectable accidents, almost undetectable, and when the Feds get too close they bump off the evidence. Murdering the engineers who engineered the accidents. Easy to do when you own your own doctors too! Make it look like a heart attack, a stroke, an asthma attack, no problem. Inject them with adrenalin voila`, a heart attack, no problem. Inject them with a massive dose of heparin, a stroke, no problem! No job too big, no job too small, if it threatens the evildoers, they will kill us all!

Need to get someone out of the picture, but not kill them, no problem. Just have your Drants switch the x-rays, so that you can make someone sick by treating them for cancer, which they don't have. If they are rich enough, you can stop the treatment and pretend to cure them, so that you can get a big donation. If they continue to be a pain, you can just keep treating them until you kill them from the toxic effects of the chemo and radiation. And the ones who are a pain, that really do have cancer, like John Gotti, no problem, just have your Drants not see the cancer, until it's too late, so that they die slow, agonizing, untreatable deaths. Need to use someone, despite an injury, no problem, just don't see it, like they didn't see Stevie Yzerman's fractured tibia, until the series was over! Unbelievable, you say, just remember who we are dealing with, the same freaks who murdered an entire generation of Canadian haemophiliac children by selling aids contaminated blood products to us. Blood products that the US government refused to buy because they knew the high risk of aids in the Arkansas prisons, so the democratic whores sold them to us, to make a buck for their campaign coffers. They could keep anybody in line with their blackmail, extortion and murder schemes. Need to make Condaleeza behave, just flash her the pictures the Russians took of her in her sleazy Frederick's of Hollywood underwear!! Need to make Ashcroft's campaign go up in ashes, send his opponent's plane down in flames to garner the sympathetic vote and put his opponent's wife in office instead, she's easier to control!

The Canadians become a pain in the ass, no problem. Just kill Trudeau's son in an avalanche that was not an accident, poison the RCMP to give them the poops in Regina, backfill manure into a well in Walkerton! That will give them something else to do, something else to think about, while we continue to poison them,

200,000 of them with the effluent from Detroit's industrial swamps.

The World Health Care Workers get too close to the truth, no problem! Collapse their scaffolding under the bridge, just like they collapsed the support under Archie's Ark! Easy to do with Charles Wasson's army corp of engineers. A few passes with a blow torch, a precision engineered razoring of a safety harness and voila` a message loud and clear for the mfer's to mind their own "bidness" and not interfere with Archie's business which was stuffing as much graft money into their campaign coffers as they could. Need to cover up the murders of the people they are bumping off in Windsor, no problem. Not when they've taken care of the local mayor's competition by putting a shotgun blast through the window of one and hiring a whore from Florida to blow the election campaign for the other. When the mayor owes them big time, covering up "little issues", are no problem!! And the police chief who likes to sample every new stripper that comes along, they have lots of pictures of him too! To keep him in line and to keep him quiet when he gets out of line. The local RCMP, no problem! Babar Baboinko remembers only what they tell him to, selective inhibition it's called. He reports what they want, like local mobsters who visit the casino, but then suddenly develops amnesia when questioned about the attempted murder of a local Lebanese patriarch. Don't ask, don't tell or you'll go to Hell, Hell Michigan that is, because they haven't the balls to say anything, do anything but what they are told like Pavlov's dogs. So the murders continue, but what they don't know is soon they will become victims too, of accidents that are not accidents as Archie, the Hords, the Pissers and the ticked off Democrats scramble to get rid of the evidence!

Now perhaps you understand why I have so much respect for the Good Guys, it took a tremendous amount of courage for them to stand up to the Psychotic Keepers of the Crypts.

My personal life during this point in time was not too pleasant either. Although I lived in an idyllic setting, right on the water in a small town north of Detroit, Hord had infiltrated there too! I found out later that the manager Penny Pith's brother worked for Hord and that they purchased the front block of condos where I lived through a trust company in Oct/97. This made surveillance very convenient. PMC occupied the 3rd floor condo above me to monitor who might visit by the harbour. The 3rd floor condo on the parking lot and lake side was occupied by Hord, so they could see who came by the lake or parking lot. The apartment directly above me was occupied by a former Contractor with Jewish mob linkages, who knew Despeisner and from whose apartment they monitored me night and day. Even in the bathroom, the f…king creeps would monitor me, because they were so afraid that people would find out what we knew all along. That Detroit, that had the highest rate of cancer in the U.S. was a f…king crypt and big business was it's keepers, and that they were poisoning us all by using the Detroit river as a sewer and killing us with the mercury, aluminum and virulent bacteria that they dumped in the river daily. No wonder our rate of cancer, birth defects and heart disease was thirty percent higher than the rest of the country. Just look across the river and think Hord and Archie first!

You did have to give them credit, they did try to kill me in innovative ways. Carbon monoxide was the first attempt, that was Charles Wasson's specialty, furnace and exhaust tampering. He even tried to use it on my youngest daughter, three months ago, the night before her law school exam. But the symptoms are easy to identify even without a CO2 monitor. Flushed face, chest tightness, headache, difficulty breathing. It didn't take a rocket scientist to figure out what they were doing and the cure was easy, you open up two windows for cross ventilation and you air the place out. When that didn't work they tried to kill my car, literally and figuratively speaking. They destroyed the engine the day before Christmas and punctured my tires so many times, I thought I was going to have to buy stock in Uniroyal. They weren't even subtle about it. Like really, how do you end up with nine dry wall screws drilled into a tire and the break-ins were funny too! They wanted to know what we knew, so they would enter my apartment at night when I left, so they could check on the progress of the cases. So I left them all scattered in huge piles around the den making it almost impossible for them to sort out what we knew. To keep track of who entered when I left because Hord's had disabled the burglar alarm, I put baby powder on all the floors by the doors. If there were dusty footprints in the hall or on the deck in the morning, I knew that our visitors had come. Then when things got really scary after they murdered my friend, I took to leaving apple cores and all kinds of things around the house. I knew that if you gave people enough to talk about, they would and they would tell me and everyone else that they were breaking in on a routine basis. When you don't have a lot of financial resources, you work with what you have and that is your intelligence!

The intelligence of a few determined people can do a tremendous amount to create balance for all the freaks sophisticated, electronic wizardry. Max Pisser's electronic wizardry. So I learned to leave my car dirty so that you could see if the doors or the trunk had been opened and I learned to leave copies of key documents and tapes in little stashes around the hospitals where I worked. I learned to pretend not to know people when I knew we were under surveillance, so they wouldn't become the next person to be skewered and I learned to watch without seeming to be watching and listen intently, so that I could understand better who in the hell were all these people and what the scope of the issues were that they were so desperate to hide. I learned to tell the good people from the bad people just from the feel of them, their aura, their presence. This was so important, because one of the biggest tricks they used in covering up their disasters was something called Silverstein's productions. They actually hired people to fake out all of the federal regulatory agents who would come to town. They would wine them and dine them and schmooze with them, pretending to be me, or one of the others that had asked for their help. Hord's press whore Swines helped with it too, wrote their lines so that the actors, the decoys knew exactly what to say to make the Feds think it was all a mistake so they would go away!

But it was hard for them to ignore us the good Docs would keep us supplied with reams of literature and pamphlets that we distributed when we picketed with the families. It was hilarious all of our handouts were being printed on Hord, Hinai and PMC's own copiers at night by the good staff that were trying to help us. They would drop them off in little bundles on our porches in the morning just like elves and the next day we would give them to the patients, so they could better protect themselves against the negligence and neglect that they were being exposed to daily during their dialysis treatments.

This pissed off the evil doers so much. The Hord whores and the Ho's of Homerica who were just muppets of Hord's owners Vanguard Index, Fidelity Magellan and the College Fund anyway. Fools who didn't give a freaking care about who they offed and why, as long as they could get an extra few cents on their dividends at the end of the month! The fools couldn't figure out who was helping us, so they started spying on me at work too and tapped my phone at home. As a matter of fact they even gave me the answering machine that I still use today, but I was told in advance that they were going to tap my calls so I thought I would make it as interesting for them as possible! They liked to jack other people around, so I thought it only fitting that we should jack them too, literally and figuratively speaking. Particularly after they murdered my friend, especially after they murdered my friend, I took every opportunity to jack them around, so we could identify and bury them, just like they did him!

Chapter 7
Mike's Almost Murder!

My brother Mike was the next casualty in their sick game to cover up the patient murders. My brother Mike was a prime target, he worked for Chrysler's and had an ex-wife who was a whore that would sell him out at the drop of a hat.

On Dec 11 the day one of the key patients Marple Fubbs died, Hord, Hinai and PMC were desperate to create a distraction and so with Archie and Prick Stein's help they arranged an accident. Mike was hit by an oncoming car at a high rate of speed. He should have died, they thought for sure they would have killed him. He suffered massive injuries, a smashed face, multiple fractured ribs and a torn aorta, the big blood vessel that leads to the heart. They took him to Windsor and the Emergency Room physicians realized right away that he would need open-heart surgery to survive. So they closed the tunnel and took him of all places to PMC's trauma hospital, who called in a cardiovascular surgeon because they didn't know what else to do. But they didn't call me because they knew exactly where I was, at Geoff Feiger's office trying to help them deal with the dialysis deaths. They knew exactly where I was because one of their flunkies was sitting outside in a city truck watching every move I made. They knew that if they called me there, I would know for sure who had caused this accident that was not an accident!

Someone in the Jewish community must have caved though, because at seven o'clock that evening, right before Mike was going in for surgery, my sister-in-law Audrey called and told me that he had been injured and probably wouldn't make it. I hopped in my car, ran down to the hospital and arrived just as Hank B. the trauma surgeon was taking him into surgery. The surgical team was wonderful and saved his life, but no matter how excellent the surgery, if follow-up care is provided by unskilled care givers whose job it is to clean the bathrooms and change sterile dressings too, then everything the surgeon has done goes to hell. That's what happened to Mike too! The first few days that he spent in the ICU were chaotic, the corporate security guards were all over me like white on rice but I was damned sure that I wasn't going to leave and let them finish him off. The docs were under tremendous pressure to get Mike the hell out of there because the hospital was expecting the Joint Commission for Hospital Accreditation to come in at any minute. They tried to extubate him, (take his breathing tube out) when he had a fever of 38.7, a heart rate of 138 and a respiratory rate of 44. I took one look at him and called Dr.B. myself and made them stop. The next day however, we weren't so lucky, I arrived after they had extubated him and sent him out to the floor, where he was still struggling to breath and had pus draining from his chest tube site. Mike had picked up the same strain of vancomycin resistant enterococcus that had killed so many of our patients. It was ironic, he survived an accident that he only had a 3% chance of recovering from and now he had VRE! So now they had the clinical pharmacist get in on the action, they knew he was severely allergic to penicillin, had almost died from a previous reaction, so they promptly hung an IV of penicillin based ampicillin and just as promptly Mike developed huge hives, swelling of his neck and throat and pulmonary edema. I walked in the door took one look at him and knew we had to get him out of there as quickly as possible or they would kill him for sure! I went out in the hall grabbed my cell phone and called the Canadian trauma team at Victorius Hospital in London to get a bed for him. Then arranged for an ambulance to come get him in the morning. I went home that night thinking that at least once back in Canada we would be safe. That was the first night Dec18/97 that I experienced Charles Wasson's handi-work with carbon monoxide. I woke up that night short of breath with a severe headache and chest tightness. One look at my bright cheeks in the mirror told me all I needed to know. They were trying to kill me with carbon monoxide. I was more angry than scared at that point, I pushed open the patio door and threw my pillow on the floor. I might freeze my ass off but I would still be alive in the morning!

The next day was a fiasco as well, they looked horrified when I arrived and soon I found out why. They had cancelled the ambulance and the bed in London! I think they figured with me dead it would be easy to finish him off. I was furious, I got a wheelchair put him in it, took him out to my car and went directly to a Windsor hospital. The emergency staff admitted him right away because he was in so much respiratory distress. But unfortunately the surgeon Cogless who saw him in Windsor was not too bright, was in Detroit's pocket or had selective inhibition. Later I learned the surgeon Cogless was in Detroit's pocket for their help in protecting many of his practice issues. After a week in hospital on conventional antibiotics, he was sent home as sick as he had come in. Twice more we brought him back to hospital with persistent high fever, rapid heart rate and respiratory failure and twice more Cogless, the surgeon couldn't see, or chose not to see the widening silhouette of his heart, the blunted angles of his lungs and the pus that drained from his chest tube site. Mike became jaundiced, confused and cold, he was in end stage septic shock from the vancomycin resistant enterococcus. So once more I put him in my car and prayed that he would survive the two hour trip to London. I had worked in their critical care trauma unit before and had seen many miracles happen there. I knew that they would see exactly what I was seeing and if anyone could save him, they could!

I wheeled him into the London E.R. and they took one look at him and did an emergency, CAT scan and echocardiogram. He had pus in the sac around his heart, which was compressing it and his lungs were filled with pus pockets. They were wonderful and took him to the thoracic ICU right away and started him on triple antibiotics. Dr. Maintainer wanted to take him to surgery but we needed to know exactly where the graft on the aorta was placed so the surgery could be performed safely. We called PMC and the f...king, rotten, piano playing corporate whores wouldn't release his chart! They, who had almost killed him three times, were going to let him die because they were afraid of their, "legal liability". At that point I figured we would give them something to really worry about. I called the Feds, I called Geoff Feiger and then I called PMC's corporate whores and told them clearly and directly that if they didn't release the chart, we were going to charge them with attempted murder. That got their attention and the chart was released!

Mike had started to get a little bit better the potent mixture of antibiotics that coursed through his veins was slowly resolving the pus that had been encroaching on his heart and collapsing his lungs. We decided to sit tight and wait, if we got lucky, really lucky maybe they wouldn't have to take him back to surgery, a second open heart surgery in less than a month that he might not recover from. So we waited and said our prayers and he got stronger every day. After three weeks on the triple antibiotics, he was well enough to go home. He needed to stay on oral antibiotics for the next six weeks at home, but he was going to survive. Now all we had to do was run interference for him against his bitch of an ex-wife who falsely accused him of threatening her. The bitch and her boyfriend, who broke into Mike's home the night that he lay dying in the emergency room at PMC and ransacked the place, took everything of value, then tried to cash his insurance policy the next day. Tried to get what they could while they could because they were in on it too! They had worked delivering newspapers for one of the Detroit papers that had been paid to cover up the story and willingly did their part too, for a little quick cash and the promise of easy insurance money. What sick sons of bitches these people were!

Unfortunately Mike's trials weren't over, now that he had survived, they had to keep him down and keep him quiet too. It had embarrassed Cogless the local surgeon, that he had screwed up so majorly, that was bad for business. So they started feeding Mike painkillers through his local GP. They gave him morphine for Christ sake, of all the non addictive painkillers they could have given him, they hooked him on morphine so that he would walk around drugged and dazed, not realizing that they had almost murdered him in their quest to cover up all the patients deaths that they had already offed!

But I knew and so did the Feds and we were just waiting for the watch to change, for the psychotic Democrats and Archie's animals to lose power. Code 88 forbids any federal agent from taking action against the current president and his government. But once the watch has changed, the big yellow hammer can come down on the right heads this time! However, this time the Yellow is the Yellow rose of Texas hammer, not the chicken shit yellow hammer of a bunch of political hacks and their ants, who knew all along that they were poisoning us with contaminated water and wouldn't do a stinking thing about it!

Chapter 8
Braveheart The First

After I knew that Mike would recover, I went back after Detroit with a vengeance. They had killed my patients, almost killed my brother and were continuing to expose the 1,200,000 people on both sides of the border to premature death from the toxic sludge that they dumped in the river daily. Where were these people heads and who in the hell did they think they were that they could bump off anyone they chose! It took a while but we did find out exactly who they were on both sides of the border and who had the goods on whom and who was in which ones pockets. I listened, I watched, I waited, I played dumb. Even when they drugged me too, the bad Drants. I made my little notes and tapes, while they all embarrassed themselves in front of the good federal agents who were waiting for the watch to change to whack the evil doers on the ass.

Braveheart the First was a federal agent who would come and see me at a little hospital where I worked in a place that Despeisner called Madison Swamp. He was there undercover, auditing the fraudulent billings that a local Health Maintenance Organization was scamming the Government for. I had seen him before in all kinds of funny places, at the 7/11, in Myers, at the gas station, so I knew that he had been following me around.

So when he plunked himself down beside me, in his wrinkly white shirt and he said, "Zes, I am the history and physical doctor". I knew exactly who he was, the one that was helping us bring, all of the kingdom down. I knew that his shirt was wrinkly because he had to live out of the trunk of his car while he did his work and I knew that if I identified him to the PMC, who had been to the hospital two days before and who had us under surveillance as we spoke, that they would kill him too! So I pretended not to know him, read the paper and ignored him, prayed that he would go away before the corporate whores identified him. But it was too late, Taresa had set him up, for a cheap lease on a Taurus. She set him, who was trying to help us, trying so hard to help all the patients, up for a cheap lease and thirty pieces of silver.

He came back six more times after that and each and every time they set him up, Taresa, Despeisner and Archie's animals. They burned his hands in a generator explosion, beat him and broke his jaw so he couldn't talk and finally shackled him to a chair and disembowelled him, just because he was trying to help. I found out later through Despeisner exactly what they did to him. I have a knack for finding people who have the answers, so when he told me exactly how they murdered him, in the same specific, graphic detail that Taresa had, I knew who killed him and why. I also found out that they knew all along that he was a federal agent, but that never stopped them. They would murder anyone, for Christ sake, they even murdered Lady Di, just because she was going to marry an Arab. Not a black, not a Jew, not a political hack but an Arab, this enraged them so, that they murdered her too and the bastards were proud of it! The sick, sick, sick bastards were proud because they had done it and gotten away with it!

The freaks were experts at high-speed accidents that were not accidents and they even planned it to occur in France, because they knew the French ambulance services treated their patient's at the scene and that with her massive injuries, in all likelihood she would not survive if she were delayed at the scene. What she would need is urgent exploratory surgery to survive. But France's system wasn't set up that way. So they gloated and they bragged because they got away with that one too! So you see they thought nothing of killing a federal agent, nothing of using electric cattle prods on them, so they would spill the beans on their leaders. Intimate details like which agents cheated on their wives and who liked to slip two fingers in and then smell them after having sex. Specific details like the fact that Louis number one had a problem with delayed ejaculation and Louis number two was just the opposite, he came with anyone and everyone, that's why the F.B.I. had his wife travel with him, to keep him from getting "distracted". The freaks wanted to know everything. They kept dossiers on everybody so that they would know how to bring down the Good Guys who were threatening to collapse the Cracker Democratic Ark, the Rabid Rabbi Jewish Group, the Jesse Jackass Coalition and all the flunkies big and small who floated in their leaky boats!

So they thought nothing of killing Braveheart the First, but I thought everything about it, all the time and I promised that since they murdered him, that we were going to bring the Mfers down brick by brick. That is how I met all the other wonderful Bravehearts, Oaktree Men and Warriors from all walks of life, who still fight the battles to keep us all safe. Safe from the psychotic hacks who would sell anyone, murder anyone in their quest for absolute power, their ego trips and thirty pieces of silver.

Chapter 9
The Bonaparte Connection

They called Detroit the Brazil of the North for all of the people that disappeared there. I learned that the first time that I had to call the local police because there were two goons in ski masks sitting in a boat right in front of where I lived. I mean give me a break, it was the end of October and they were the only boat on the whole lake and they had parked themselves thirty feet from where I lived. I went out on the deck, snapped their picture in case they disappeared and called the police. The police were very nice but very clear. They said that they had several suspicious deaths related to people who were in conflict with a couple of the large Detroit health care systems and suggested that I leave as soon as possible. I couldn't do that however, first of all we weren't done yet, we hadn't closed Hinai or protected the patients in the outpatient units, so I stayed and I learned so much more!

It was very, very, easy to make dissidents disappear in the city of Detroit, especially when you were the head of the police farce. When people came to inquire about their missing loved ones, the farce didn't have to look, they knew right where 90% of them were, cut up in small packages, in Bonaparte size packages for handy burial at the local body repository, the moth-balled Hewlett Packard plant. But the locals knew, day and night they would see squad cars, city of Detroit service cars and dump trucks, pulling into the plant to dispose of more packages. The locals knew too, that if they went to the police farce to express concern about a murder, that they were likely to end up in jail themselves as a "witness", until the farce had intimidated them into silence. The local's knew that they couldn't go to the state attorney general's office either because they probably would end up in the body repository too, for their efforts! Hell you could even hang world health care workers from the bridge and their chief prosecutor Bob Assani would look in the other direction because he already knew who did it and why.

Some of the local FBI were questionable too, they liked the free drugs and fast women supplied to them by Detroit's finest in exchange for being blind mice. Seeing nothing, hearing nothing, knowing nothing, blocking the flow of information to Louis Freeh, St. Louis, the guy that Despeisner called free Willy because they worked him like a dog and so cheaply. They only paid him $94,000 a year, used his brilliance for $94,000 a year and deeply despised him because he couldn't be bought. But Louis had a flaw too, he trusted the farces implicitly, his own and Benny's and for that they made a mockery of him, jerked Louis Freeh around too, setting him up in staged decoy operations that were orchestrated by Bonaparte, Snalexander and Sweins. Silverstein's productions they were called, designed to fake out the good FBI, so that they would go away and leave Bonaparte and his jets alone! But something happened during the OAS. The OA f...king S, where the leaders of the free world, including Blow Job Bill would hold their economic summit. As they checked potential sites they saw clearly that they could never hold the conference in Detroit. There was no police support, no fire services, no emergency services, the water supply had so much shit in it, that you could almost walk on it and the power supply was sketchy at best. And heaven forbid if they had to rely on Fermi 11 for additional power, it would blow them all literally and figuratively to kingdom come! Then they checked the two major hospital systems they had to rely on downtown for support and soon realized why the locals called them freaking crypts. They had more junkies per acre on staff than Vancouver Park and a mortality rate that would make the Bronx zoo look like the Mayo Clinic! They had no choice, no freaking choice, but to swallow their American pride and hold the conference in Windsor. There it would be easier to handle the small time crooks that ran the city and control some of the human debris that floated through the bridge and tunnel on a daily basis.

The Packard Plant, the good FBI didn't know what to do about that! They had told Janet Reno, but she wouldn't touch it with a ten-foot pole. Clinton needed Archie's votes and she had already been burned over the Lewinsky thing. Reno was a political animal and so she backed away from that mess as fast as her flatties would carry her, leaving Louis to sort that one out for himself and he did! So Louis called his friends at The Hague in the Netherlands, who specialized in the excavation and identification of mass grave sites to come in and clean it up and shut it down. What a fight that was! Benny had swat teams patrolling the perimeter like they were guarding the crown jewels and Charles Wasson helped too, they brought in junkies to hold all night raves in the building and supplied them with barrels of contaminated waste that they could burn and toss around. Just so that Wasson could have some of his army corp of engineer friends who now worked for the Department of Environmental Quality, come in and declare it a contaminated waste site, giving them a reason to tear the building down and hide the bodies forever. But the police farce gave themselves away, guarding the place day and night, fighting the Hague officials tooth and nail to block their access and the more the police fought, the more the FBI knew that the rumours were all true, that there were over 800 neat, small, square packages with Benny and Archie's names all over them, of all the people who had disappeared from the Brazil of the North.

So Archie's animals true to form, to get even with the Hague for meddling in their business, sent a New Year's Eve present to them, to the Hague to spoil their New Year, just like the Dutch had spoiled Benny's. They sent a partier, a representative from their fire department, who lit a sparkler and ran like hell, out the doors that Benny's boys locked behind them. Leaving hundreds of Dutch partiers to literally and figuratively burn to death, just to give the Dutch a message not to fuck with big "bidness" in Detroit!

Now, perhaps you see why I call them animals.

Chapter 10: Max Pisser the Pimp & the Rabid Rabbis

Whenever I think of the evilness of the radical Jews, I try to balance it with thoughts of the good people like Lathan Nevine who tried, tries so hard to do the right thing and take good care of people and Hassi Brickman who was a truly saintly woman who lived her values daily, quietly, peacefully, brilliantly. There were so many evil Jews however, the rabid fundamentalist Rabbi sect, who arranged the murder of Lady Di because she was going to marry an Arab, had the audacity to marry an Arab, after they had sent her all their finest and she refused them. Refused to marry into a group who raped, murdered and destroyed the Palestinians, just because they were Palestinians. Racists to the Nth degree, the Orthodox Jews played a continual game of chess. Oh they allowed intermarriage alright but it had to be to the right people. So they sent their Fannies and their Frannies for extensive plastic surgery in order to make them pretty, so they could marry them off to any race or creed as long as they had money. Money that could be funnelled into their coffers and war chests to make Israel win the holy war! What a joke that was too, there was no holiness to the war, it was all about money, power and control and a good excuse to bilk the community! Max Pisser, Don Schmancy and the Rabid Rabbis sold off anything and everyone they could for a buck, power and control. Max Pisser the Phyllo-anthropist who gave it to everyone in the ass. Who double crossed everyone, took everyone else's money for the Jewish cause, put two thirds in his own coffer and then gave the rest away under his name, to make himself look like a hero! Max Pisser who screwed everyone literally and figuratively speaking, those he couldn't pimp off on his HIV+ offspring the virgin Scarey, he screwed literally through destroying them personally and professionally. Al Raubman was a target of his. Al who was so funny, that he had the audacity to rob from the rich, by jacking up the prices at Motheby and Cookie's. Mossad Maxie, his boy Goings and the Homerica whores fixed him with Despeisners' help. They set him up in a government sting that would pour more soft money into the democratic war chests and embarrass Al at the same time. It was easy to do, just throw a couple of bucks at the whores who ran the New York office and the Fairlane Mall and voila` they skewered Al too! Then to finish him off, to make sure what remained of his money went in the right direction, their direction, they hooked him up with a former Miss Israel who could snuff him at a moments notice if there was a need! Max Pisser liked Archie and Benny, the ngrants he called them, running the city of Troit because they would do anything for a few bucks thrown into their own war chest too! Don Schmancy was Maxie's sidekick, the Don, the jewish Don who had never had a good hair day in his life! Don was so tacky that he would pay to have a woman's hair fried because they wouldn't have anything to do with him! Him with his hundreds of millions of billions of dollars, because he reeked of what he was, a shyster, a shylock, a shrew, a crone to the nth degree. Who was so cheap that he took his own wife to a discount plastic surgeon to have her eyes done, to make her look a little bit better and the sutures fell out, literally and figuratively speaking and on purposely speaking. This was the plastic surgeons' revenge. They could be such a funny group too! In Royal Oak, in their utter disdain for all the Fannies and Frannies in their jewelled flatties, the plastic surgeons gave them all the same nose! The same nose to go with the same blonde dye job and the black turtlenecks they all wore to hide their surgical scars! Shopping in Royal Oak was a hoot, you thought you were in Stepford, with a blonde clone at every turn and a cashmere-coated shyster at her side. Shyster was the term for a Jewish lawyer and they were good at it too! They played the piano at whatever key and pitch you would like to hear, as long as it made them a buck! It didn't matter who died or how many died, they defended whoever, whatever, whenever for their thirty pieces of silver. When there were no problems, they helped create them. Like helping to plan the back filling of manure into the Walkerton well, so they could be perceived as Heroes' protecting the little people, while they cleaned up and raped the government and the little people for millions of pieces of silver. Don't ask, don't tell or you'll go to Hell, where they all are already. Planning to continue to rape, murder and kill, anyone or anything for a buck!

The jews of the massed muppet gang were great buyers and sellers of information, using their board connections with the phone companies, the police departments and the banks, to illegally tap phones and personal files to get information. To get the scoop on whoever was threatening the Jewish National Trust Fund! Their motto was, "we take good care of our own and all others we take care of!" Literally and figuratively speaking, stopping at nothing arranging murders, rapes, stabbings, whatever it takes to deal with those who threaten their ability to win the holy war, the holy grail, win the geese with the golden eggs! So you can see why we were not their favourite people. In exposing Hinai for what it was and in exposing Pisser and the Rabid Rabbis for who they really were, freaks who would allow the lives of 1,200,000 people to be compromised to save a few bucks. We became targets of the jewish mossad muppet community too! All because we wouldn't lie down and let them murder us all!

But God did a funny there too! The Jews demanded a sacrificial lamb for the Hord's complicity in this mess! So poor boy Hord, Hay Hord Jr. was auctioned off to one of the clones in her flatties. The torah, torah, torah came home and Hay Hord Jr. lived unhappily ever after! Now how is that for poetic justice!

Chapter 11
The Words Get in the Way!

The newspapers were a problem from the very start, at least one of them! The Detroit Metro Times was wonderful at first, they had the guts to print the story in the beginning but backed off later when their black editor realized that Archie and the black community were as much at fault as the Hords and the bad Jews. Yes Hord and Hinai delivered the final product water that killed our patients, but the shit stew that came in from the street all belonged to Archie! The Free Press was a write off as well, they printed whatever big business told them to in typical lemming fashion in order to preserve Hord's big ad accounts. They adopted the attitude of see no evil, hear no evil, speak no evil in order to get the bulk of the Hord and Jewish "bidness".

The Detroit News was great and boy how the evil doers hated them. They said no news was good news and they would actually raid the news stands to get all the copies, so people couldn't read them. They even paid the people who were supposed to deliver both papers under the joint operating agreement not to deliver the News because they had the balls to print the truth! When that didn't work, they threatened the reporters too, slashing their tires and hiring hit men like Moxley from Boston, the Tratteneri's from New York, the Boycano Brigade from the Philippine's and the Ivory gang from Jamaica to bump them off. Claxton and Hurt were specific targets. Do you want to see Claxton get creamed they said. Do you want to see Hurt get hurt they said I mean really, why would an organization like Hord's hire a hit man like Moxley to do "contract" work for them. Contract work my ass, more like concrete work, to plant Claxton and Hurt in pillars on the Hord freeway!

Prick Stein and Wasson, two more mossad muppets and semi-colon puppets were the same city administrators who sent their firemen into unsafe burning buildings with nothing to work with. Buildings that they, the city had built with their own CONstruction companies. Knowingly and wilfully built far below code, just so they could torch them, collect the insurance and get full market value for the buildings. Buildings that were really just a shell, built that way on purpose. The freaking bastards would send their own firemen in to fight these fires knowing that the floors would collapse under their weight, to keep them in line too! Archie's line, Stein's line, Charles' Wasson's line. Literally hundreds of fire fighters were injured fighting the Bunker's CONstruction company fires and the city didn't care, didn't give a flying f...k just as long as their coffers were full. When the lives of 1,000,000 people were at stake in the hospitals, the dialysis units, the nursing homes, even those who sat on the top floors of the ivory towers downtown, when their lives were at stake because the firemen had nothing to fight the fires with, the animals in Archie's bunkers didn't care, their coffers were full. So instead Archie and his animals threatened the lives of the reporters who tried to protect the people. Bullied and intimidated them into silence, so that they could continue their charade, their deathly façade of a city that was a house of cards. Tony had said Archie the Zero diddled while Rome burned and he was right!

Chapter 12
Snowman Hung and the Jesse Jackass Coalition

You would have thought if anything that the black community would have rallied to protect our predominantly afro-American patient population, but what a freaking side show that turned out to be! The Jews were nothing in comparison to the Black elite's run for the trough! The running of the bulls at Pampalona was nothing in comparison to the tap dancing we saw to protect the sintegrity of the black community. We went to everybody, Snowman Hung's cousin Clod, Smelvin Durfy, Whoryln Cheeks Schillpatrick, Schillman Gary, the Southern Christian Leadership Council, every freaking scion in the black community trying to get help. But they were too busy to help us stop the carnage, to address the contaminated water issues, the fire safety issues and the formaldehyde issues in the "die-alysis" units. Too busy at the trough of free social lunches celebrating their monuments to bowel movements. Too busy to give a flying f...k about the fact that Detroit had a higher mortality rate than a third world country, a dialysis death rate 25 to 50% higher than Canada, France or Germany and that their own people were dying like flies. Coretta Snot King what a joke and Norris and the black nephrology leadership council, what a joke. They all had partial ownership in the same units that were pickling the patients with formaldehyde. Al Sheepton, what a joke, Al who came flying down to crucify a security guard who had did in one of their "finest". So ready to attack Raubman's store, then he wouldn't even back down when they found out the security guard was black too, because he figured there was a buck to be had in it. Jesse Jackass, he was too busy too, literally and figuratively speaking, screwing any nubile young lovely who crossed his path! Jesse was great at the laying of hands he laid his hands all over any one who would let him! Jesse was too busy extorting millions of dollars in protection money from companies like AT&T and Anheiuser Busch, so that he and his boys would become millionaires by extorting lucrative beer dealerships from some wussy companies who were afraid that they would get hit with one of the sham discrimination suits that Jackson was so great in setting up. Semi-colonel forget it, he was too busy poisoning his own troops with poison gas in Desert Storm, too busy vacationing in perfect silence on Lake Powell. The chief corporate counsel at Chrysler's and their colleagues at Schmoois and Stunday, too busy helping Schmeaton strip the company before the German's took over. The director's of perversity at GM, U of M and the hospitals, forget it. They were too busy covering their tracks that glowed in the dark like the contaminated waste in the Detroit River. Too busy because in the words of Rhett Butler, quite frankly my dear, they didn't give a damn. They were comfortable and their coffers were full, so they weren't going to rock the boat. So they turned their backs on their own people and all the people who were being exposed to the toxic sludge daily. 1,200,000 people and for three more years the carnage continued. Affirmative action and diversity, what a joke, the only affirmative action they were interested in was what they could do for themselves. They subscribed to semi-colonel's (Powell's) motto, "Take care of yourself first"!

The other freak in the black sand pile was Artass Whortring, a Drant flunky from Sierra Leone, who turned in the Pink bandit who was helping us, so that he could take his job as CEO of PMC. The freak who makes megabucks, treating the cancers that Detroit is causing and pickling the babies and children on dialysis at PMC's own children's hospital with formaldehyde. The freak that interfered with my brother's care in Detroit, Windsor and London, so PMC wouldn't "look" bad. The freak who likes to arrange the drugging of nurses for a joke, to protect any and all physician malpractice. But the joke is on him now, Artass and his central porter ring may find himself with a new address, but in Leavenworth, not Washington, Tadum! Oh, and I almost forgot Drant Uish, whose "Uish" it is for everyone to get environmentally-caused cancers, to make him rich as CEO of the Michigan cancer program.

But there is a God too; one in every six afro-Americans will end up on, "die-alysis". Each one will be subjected to formaldehyde every treatment, experiencing the same substandard care and being at risk to catch one of the super infections every time they have to go in the hospital and have their access revised. One in every six of them will be subjected to the same horrors that our patients both black and white, had to live through daily. One in six, six, six and I truly hope God has a sense of humour and makes sure that Archie, Charles Wasson, Jesse Jackass, Al Sheepton,

Artass Whortring, Powell and the rest of their whores are standing in that number six check out line!

That would be the greatest poetic "justass" of it all!

Chapter 13
The Piano Players

How fitting that chapter 13 end up being the piano players! I probably know a thousand of them by now and of that thousand I can honestly say that three are decent human beings! I thought about being a piano player once, but I couldn't stomach the conversion and the more I met, the more I understood the joke, "What do you call a corporate attorney?" Answer: " A piano player in a whore house!" My God what a freaking sick, psychotic group they are. No wonder Tom Monaghan went down river and started his own law school! If any profession needed an AVE MARIE, that one did! It's probably best to reference them in sequence, so let's start with the little jarhead first! Despeisner called the first attorney we picked, the mouse that roared because he made a lot of noise, but he didn't do his homework. It was very easy for them to buy him off, first of all his wife was trying to get pregnant, so Hord threw a bunch of free fertility treatments at them and secondly his wife worked for a predominantly jewish law firm too, who wasn't going to do anything to ruffle the Jewish community's feathers. Snopatin and Schiller where the jarhead's wife worked was the next firm that we dealt with, they were more than happy to go after Hord, but wouldn't touch Hinai with a ten-foot pole. They caved too as soon as Hord's Schnellburg helped them to understand that they were as complicit as Hord now in the patients' deaths. Schnellburg, Snellen and Honberg what a trio that was, corporate whores to the nth degree who arranged all the meetings to edit the charts and get the players to rehearse their lines so that they knew exactly what to say and what not to say when the regulatory authorities came in. The same corporate whores who repeatedly parroted that, " We would do nothing to trigger an investigation", despite the fact the law very clearly said what we must report!

The same corporate whores that sent letters to the patient's and press releases to the papers saying there is no problem, while they worked furiously until three A.M. in the morning altering charts and covering up the evidence that showed they were as guilty as Hell. Hell, Michigan, that's what they called it themselves and they called the hospitals the crypts because they knew exactly what went on there, knew, because they were the one's covering it all up!

I did meet one good lawyer at this time though. The contracted attorney who had dealt with the baby deaths from contaminated water at one of PMC's other hospitals a few months before. He was great, he truly cared and truly did try to get them to do the right thing. So the bastards fired him from the case and from his firm, Scillmouth and Hanury. Fired him because he had a conscience and did care, the bastards! So now you see why I took my little tape recorder with me wherever I went, it was because I knew exactly how these animals played the game and what crooked, evil sons of bitches they were!

The next bunch we had to deal with were some freaks from Thomas Lyer and ASSociates who had the audacity to lie in front of a federal court judge two weeks before the report that crucified them, came out from the Health Care Financing Office in Chicago! That's why they set him up with Jac the Knife's wife, so they would have something to hold over him when the case came to court. Av Stones, who wouldn't have touched her with a ten foot pole, if they hadn't drugged him, she was a real triple bagger! But hell drugging a judge was no problem for the sansibelt Hords they had fed Jimmy Hoffa through a meat grinder for Christ sake. Do you think a little thing like a federal judge was going to stop them?

Then we inherited a few more freaks from Schmoois and Stunday, the same firm who were on Homerica's mossad board of trustees with Hord's asshole windbag, Snale Horden .The piano players at Stunday's were so stupid that they would actually have documents faxed to my home that they had prepared at 3:00 am in the morning! This let us know for sure that we had them by the raisins big time. My God they were salivating, just salivating to find out who was helping us so they could murder them too! So I fixed them, I figured since they could lie to a federal court judge, I could lie to them! So I told them I was pregnant, pretended I was pregnant so we could get them to stop using formaldehyde and close Hinai down without going to court, so I wouldn't be forced to spill the beans on all the people that had helped us. I even dressed up in a little black velvet maternity dress, trotted down to Orchestra Hall, PMC's corporate headquarters and pretended to be willing to help them, just so they would think the pregnancy was true. And they fell for it, the stupid dummies fell for it, they were so used to lying to everyone else that they didn't know they were being had! Getting screwed without a kiss, giving us what we needed to make our patients' safe without it costing us a cent and without telling them a thing! Not bad for a little team with no resources! We knew all along that Scott Crombs was in with Fucilli's boys and that they were going to set up the bad World Health Organization guy with Schmamy Harkman. We thought that was fitting, he was here to protect the patient's and he sold out to Hords, big banks and big business. So they lined him up with Schmamy, who had a history of setting her previous husband's beds on fire. We figured that would give him something else to worry about. Hard to focus on screwing the good World Health Care People when he had to bring a fire extinguisher to bed with him every night! We giggled privately about having given it to Max Pisser, Don Schmancy, the Hords, Hinai, PMC, Archie's Bunker and the Cracker Democrats. We had a little acronym that we used, BOHICA that stood for, "Bend Over Here is Canada"; and our efforts did protect our American friends for a little while, until Taresa sold them out, gave everything away for a cheap lease on a Taurus and a few bucks. Once she did that and they murdered my friend, I vowed that we would bring them down, someway, somehow, and now we have, we are bringing the mfers down, simply because they didn't know when to lay down. They bragged and they gloated and made asses of themselves in front of the whole world. Letting everyone know that they did kill doctors to keep them silent and they did drug nurses to keep them silent and arranged accidents that were not accidents for everyone that tried to help! They thought they could get away with it, they had everyone in on it, even the former f...king president of the United States. Billy was so worried about losing the black vote and his press secretary John Franssen was so worried about the truth getting out that they involved themselves in the cover-up. That was their downfall, Flloyd Snemple told them to keep it simple, but they wouldn't listen. As a result, they lost and now the big Santa, George Herbert Walker Bush Sr. and his attorney general John Ashcroft are making a list and checking it twice. For the democrats there won't be a next time around in the white house. Not once everyone knows that they helped cover up the murders or our patients and the systematic poisoning of the 1,200,000 people who live here on both sides of the border. The same democrats who set –up 900 F.B.I. officers for exposure, in their last days of office. This time the keepers of the crypts screwed up big time and the Good Guys are out to get them! Doing "bidness" in the crypts is never going to be the same again! Ask Bank First Suisse of Boston one of the owners of Hord's who has lost 818 million dollars this quarter or Hord's who just took another 3 billion dollar hit on the tires that 8 months ago, they said there were no problems with, despite the fact that people were dying like flies. Ask them if they like our way of doing "bidness". Like Marion Berry and Archie got hung by their own vines, the democrats will have to walk a new line, one to the federal pen and the bankruptcy lines! Tadum!

The Piano Players: Part 2

The Canadian piano players weren't much better! The first jerk that we had to contend with was a fool named Clouessa who was worried about a relative, who was doing a surgical residency in Mount Clemens. A twit who had almost done in some of his patients because of his gross incompetence. He was so bad in fact, that he was in danger of being thrown out of the residency program and he hated me because when I worked in Mount Clemens, I wouldn't let him do things that would harm the patients. I would refuse to carry out his orders and call the attendings, who always sided with me. This pissed him off so much that he ran home and cried to relatives, who with Despeisner's help tried to frame me over here too!

Then we had Depuca, protector of the polish honour. Polish honour my ass! I knew a few good poles, but the ones that tried to cover up the contaminated water issues just to keep Hord's business, certainly weren't included in that group. The poles who were poisoning their own children in Hamtramck with lead levels in their water five times the safe limit. Something that they and the state had known about since 1992, but were too freaking cheap to fix! Do you think they gave a flying f...k about the Canadian children when they didn't even care that their own kids were suffering from brain damage from the lead in the pipes? Do you think Depuca and his piano playing relatives cared about the Canadian children dying from cancer caused by the PCB's and heavy metal toxins in the water, as long as he could keep on the good side of Hord's, no f...king way! He'd kill his own grandchildren to make a buck and his wife who sat on her hot air filled ass on the police commission board, passing judgement on things they knew absolutely nothing about and were too stupid to research. So instead they relied on what Archie and Benny and Wasson and the boys told them and as a result got screwed royally, because now they were as complicit in the deaths from this very real Watergate as their Detroit colleagues were.

Depuca said they didn't like my mafia affiliation, it offended their sensibilities! What a joke that was, I would not still be alive if it wasn't for the good Mafia people who protected me from the bastards who ran the crypts in Detroit. Every day, the good Mafia people, the Italian Mafia people would come and check to see that I was okay. They knew that the manager of the building where I lived was on the take from Hord and they knew about the bad people who lived above me. They even knew about the few bad Drants at work who tried to drug me. For over a year they protected me, coming to tend the yard in front of where I lived every morning, checking to see if I was okay. But finally they were infiltrated too, one morning when I came home, there was a different face on the crew, a face that I knew from Detroit and I realized that we had trouble there too! But back to Depuca, when he tried to pull his holier than thou crap on me. When he knew as well as I did about all the crap that was going on on both sides of the border, I just blew him off! Maybe he would like to explain his polish honour to our dead patients or my friend who was murdered or to local philanthropists Tony Toldo, Tom Joy and Leon Paroian whose prostate cancers were probably caused by all of the mercury in the water. Explain why a tiny city like Windsor has to build a freaking hospice village to accommodate all the people that are dying like flies here. Explain his honour to them, and then come back and see me! Asshole!

Then we had the local mayor, whose competition had been blown away by the boys from the crypts, literally and figuratively speaking. They put a shotgun blast through the front window of one and hired a whore from Florida to blow the campaign of the other. It was much easier to do "bidness" with someone you already had in your pocket. But Archie's semi-colon puppet gang was screwing Mikey the Windsor mossad muppet over too! Systematically, stealing back the casino business, the auto business, even the health care business. Archie shook his hand with one hand and stole Windsor's revenue with the other. Stole the nurse's who used to staff the Canadian Hospitals, to staff their crypts because no Americans would work for them and used Windsor as a playground for their out of control youth. Better that they trash Windsor, than disrupt Archie's lucrative gambling big "bidness" downtown!

Then of course we have the Cogless Clan, whose surgeon brother almost killed mine through his incompetence and who was systematically collapsing a Windsor hospital through his stupidity. They built the new addition on the hospital without consulting the physicians and then were shocked to find that they would have to tear half the place down to get the x-ray equipment in! But they had a plan to cover their ass there too, with Despeisner's help. They just visited Hord's Kingswood warehouse that was so full of mould, that it's walls were green. Scraped off a little mould and then blew it all over the walls of the Windsor hospital's new addition. That way they could get the place torn down and rebuilt at little or no cost to them and delay the project until the x-ray equipment that had never been ordered could be brought in for installation, before the project could resume again! The piano player Cogless who had been begged repeatedly to hire decent security staff to protect the nurses in the ER and on the psych unit, who said," no, there is no problem," until one of the nurse's was almost beaten to death, again! Cogless the piano player who compromised the lives of five severely injured trauma patients by delaying a surgeon at a local hockey game because they were winning. No job too big, no job too small, the psychotic fools from the crypts and the local flunkies they have on a leash, do it all! They even helped cover up the hanging of the World Health Care workers from the bridge, that is until the WHO became as mad as cows about it and sent a little message to Hotel Don't of their own. I think that Hotel Don't learned a very expensive lesson that being complicit in mossad homicides gets you into a lot of hot water. Steam sterilizing hot water, to make up for covering up the contaminated water that they were all so desperately trying to ignore! So desperate in fact that Mikey had to bring his relatives to town to work at the local hospitals to help cover up all the deaths. But Mikey's plan won't work, covering up the killing fields of Windsor is finished, because the good guys have their number and the "fambly's" of the bad guys are going to be decimated until they tell the truth and fix the problems. Haven't you noticed the number of deaths in the local whore population? I have and it will continue until the cryptic clowns straighten up tell the truth and fly right. Maybe then their string of bad luck of car accidents, falls, suicides, homicides and surgical misadventures will end.

Then of course we have Harvey and his disappearing rabbits and his boy Charmin. Harvey hadn't been very happy about losing all of his rabbits, his "bidness" to his boy Charmin. So the good old boys of the crypts decided to make a little "bidness" for him in exchange for making sure that his boy Charmin didn't look too closely into the death and the attempted murder of the World Health Care workers they hung from the bridge. So the democratic boys from the crypts went up north and backfilled some cow shit into a well. They picked Walkerton as a joke, get it, they were shitting on George Herbert Walker Bush, because he tried to help save all the people here too! So they went and backfilled cowshit into a well, but no one was supposed to die. It was supposed to be a joke, like when they gave the poops to the 600 R.C.M.P. officers in Regina. It was supposed to make people sick, to generate a little "bidness" for mossad muppet Harvey, make him a hero. It wasn't supposed to kill them, but low and behold, another city administrator and another Drant stepped in and tried to cover it up, just like they did in Detroit. Seven people died and 2300 were sickened, because a bunch of psychotic freaks were trying to create a diversion from the mess that they had in Detroit and protect big "bidness", who is systematically killing us all with the shit and toxins they dump in the Detroit River daily. Not very pretty is it.

Now perhaps you understand why I was so upset when I first came home. My God, what animals, no conscience, no compassion, no remorse. They would kill anyone for power, control, their ego trips and thirty pieces of silver. What a bunch of f...king freaks!

Chapter 14
The Marple Fubbs Fiasco

The Marple Fubbs case was one of the ones that the piano players desperately wanted to hide. Poor Marple, she had survived having twelve kids and three husbands, but she was no match for the systematic negligence that she was exposed to daily during her dialysis treatments, the negligence which finally killed her. I first learned of the Marple Fubbs case in the spring of '98, through a contact at the Detroit Metro Times. She was another patient who had gotten ill every time she was "die-alyzed" in Hinai hospital, but we rapidly found out that that wasn't the only issue that led to her demise. What she went through would even have killed Bull Durham too and probably a lot sooner. Shortly after the Metro Times had called me, the Fubbs family called and asked if I would help them find out what really happened to their mother. I did and ever since that time, I've had to play a constant chess game with the freaks to stay alive and protect the Good Guys. We used to joke about ending up in a pillar on the Hord freeway. I would tell them to look for the one with the little red high heels sticking out and then honk and wave, because they would know that one would be me! But they said that Hord would never be content to put me in just one pillar and boy were they right!

It was about this time when I was commuting back and forth to Detroit to help the family's picket, that I first noticed the trucks. The Hord, PMC, Hinai and the city of Detroit goons were ever present, primarily following me around in an array of Hord cars. But then a funny thing started to happen, the truckers began running interference. It became like a football game with the Lions always losing! It was fun, Hord would cut me off or sit on my tail and then a great big transport truck would come wheeling in between us and I would go merrily on my way, safe between the bumpers of two big trucks! I am sure they were wondering what in the world this was all about and why were the keepers of the crypts trying so hard to kill me, literally and figuratively speaking. So they became my tail too, but a welcome one because they would create balance for the goons in the mercury's who were so concerned that we would expose where all the mercury tailings were coming from in the river. The mercury tailings that were causing our rate of cancer and birth defects to be thirty percent higher here than anywhere else in Canada . The mercury tailings from their ignition switches, that they dumped in the river daily!

Sorry to digress, we will get back to the truckers later in the chapters on the Good Guys. The sentinel event that did Marple in was an accidental formaldehyde bolus that was caused by a brand new nurse. A little local Canadian brand new nurse that they ended up forcing to marry one of the creeps from the crypts in order to keep things silent! The sad part in this whole scenario was that it wasn't her fault. She was poorly trained yes, but she did do the test to make sure the formaldehyde was gone from the artificial kidney. She did it, while many others around her never bothered. But what she didn't know was the chemical that she used to do the test had expired three months earlier and wouldn't have shown a positive result even if there was pure formaldehyde in the artificial kidney. What she also didn't know, was the person who processed that kidney had been hired the day before and had no training either, no training into a highly complex, exacting process that was meant to sterilize the kidneys. The third thing that she didn't know was that in all likelihood, judging from the severity of Marples reaction, there was probably almost pure formaldehyde in the die-alyzer, because there was no test kit for the reprocessing tech to check how much he had put in it. The test that was supposed to be done on every single batch of "die-alyzers", each and every day and they hadn't had any testing solution for over a year.

You want to know how we knew, because after the state came in and said there were no problems, despite the fact that the patients were dying like flies, we went to plan B. We sent all the documentation to the Federal Health Care Financing Administration in Chicago, who saw exactly what we did and came in and shut the bastards down. The bastards were furious because it cut into their bottom line by eight million dollars a year. So they sacrificed Shelley, a little white Canadian nurse who had received no training, she was A-framed, Archer framed, because she was white and Canadian. Poor Shelley, her life has been ruined too, because she is being forced to marry one of the animals from the crypts for a murder she didn't commit. All because she did lie down! Now you can see why some of our locals are looking a little sheepish these days. Shelley's innocence is all there in black and white in the report from the Health Care Financing Administration, there for them all to read if they had done a little homework! Leaders in education my ass, the only thing they are leaders in is blackmail, bribery, theft, corruption and murder! They make the Mafia look like Mother Theresa in comparison and they get away with it, that is they used to get away with it, until the new big yellow hammer, the yellow rose of Texas hammer came down! Tadum!

Sorry I digressed again, back to Marple, after Marple was taken to Hinai, a whole series of mysterious unexplained events occurred. The good news was that the Emergency room physicians did treat her appropriately for the anaphylactic shock from the formaldehyde despite the fact that Hord sent them a falsified report to say the formaldehyde test was negative. The accident happened at seven in the morning and Hord reported a negative result to them at 11:00 AM from a reference lab in New Jersey. Give me a freaking break, what did they do, charter the Concorde to get it there!

Marple's symptoms were that of a classic formaldehyde exposure, burning sensation all over her body, rapid heart rate, acute respiratory failure, high potassium level and red blood cells in her urine from the tissue destruction. The emergency room physician Dr. Pumpkinson's first note clearly detailed what had happened, the note that the freaks from Alter road made him alter the next day to say there was no problem. But we had that original note too and so did the Feds.

Then they sent Snoozin Whoreld in, what a joke of a general practitioner she was, no wonder Detroit's census dropped so rapidly, that Michigan lost a senate seat. She and her affirmative action cohorts were one of the primary reasons that Detroit's mortality rate was and still is higher than a third world country. Snoozin's job was to help cover up the black liability, so true to form they wanted to spread the complicity around a little bit, so she brought the mossad muppet cardiologists in. Now I'm sure Marple did have a cardiac event, when you look at all the research done by the University of Edmonton in Calgary it clearly shows the linkage between formaldehyde exposure and heart disease. Even for a layman this makes sense, remember how tough and leathery the tissue of the frogs were, that you used to dissect in science class! That is the same thing it does to a live person's skin and blood vessels, they shrink and become narrow and rough inside, blocking blood flow and causing the heart to be damaged. So they called the cardiologists in and that was an interesting dynamic too! She had two angiograms, two days apart that showed totally different results. One showed severe disease in only one of the main blood vessels supplying the heart, which would certainly not justify doing open-heart surgery. However, another one was done two days later by a different mossy cardiologist with linkages to Hord. The angiogram done by Dr. Cowman, showed severe triple vessel disease, which would justify doing open-heart surgery, what a miracle! Now remember all these events were occurring at the time that the keepers of the crypts were desperately trying to cover up the patient deaths that had already occurred at Hinai through the spring and summer. So they were desperate to call Marple's problem anything other than what it really was, gross negligence, but not on the part of Shelley, on the part of the freaking keepers of the crypts who wouldn't spend a few cents to do things properly. Think Hord first my ass, only in Wayne County Morgue!

So the two flunkie, mossad cardiologists Drants Cowman and Prasek recommended that Marple have immediate surgery, despite the fact that she had already picked up an infection at Hinai. She had been there less than a week and still had a fever even though she was on three very powerful antibiotics. But the dummies took her to surgery anyway so eager were they to cover the crypts tracks, they took her and in doing so, initiated the beginning of the end for Marple Fubbs! Marple's ensuing clinical course was truly like a tale from the crypts. They started her on aspirin prior to surgery, aspirin to which she had an allergy and as a result she bled like a stuck pig after the surgery. This resulted in the need to give her blood transfusions, which promptly put her heart and lungs into congestive heart failure. So they gave her nitro-glycerine, which she was allergic to and rapidly dropped her blood pressure into the toilet. But that didn't stop them they pulled her breathing tube anyway, despite the fact that she had seventeen extra pounds of fluid in her system which was systematically drowning her. So they had to "die-alyze" her to remove the extra fluids from her lungs and even though we had already met with Drant Hergsman, even though they knew all about the contaminated water cultures and the high endotoxin levels and the grossly unsafe limits of the chemical contaminants, not a freaking thing was done. They had checked nothing, corrected nothing, the a...holes! The next day, poor Marple was noted to be weeping serous drainage from her legs from the infection that had already colonized in her system, the super infection which would eventually kill her.

By Oct.13, her heart rate was up to 132, her blood pressure was down to 87/48, she had to sit up constantly because she couldn't breathe, her legs were saturating her dressings three times each shift and pus had started to drain from her chest incision. The notes from her Drants said she was for possible discharge home the next day, they were sending her home to get rid of the evidence. Better that she die at home, than she be another death from a hospital acquired infection on the books.

On that same day, Hord's Schmancy Boobola and her corporate whores had signed Marple's name to a care plan that Marple couldn't agree to, because Marple had never seen it. Marple was in the hospital that day, while the same outpatient unit that had almost killed her, was forging her name to care plans, consents and all kinds of legal documents. But the stupid a...holes were true to form, they forged her name but misspelled both her first and last names. We know this too because we have all those copies and we gave them to the Feds as well! This message is for Chez Fendry, the next time you forge someone's name, at least have the courtesy to spell it right! Assholes!

At the same time that they were offing poor Marple in the "die-alysis" unit upstairs, the Director of Hinai's infection control services and I were reviewing all of the bacterial culture results and endotoxin levels from prior years as well. So many were missing that we went down to the lab to find out where they were and low and behold, who was in the lab but a Hord "die-alysis" nurse Schmildred Stavarro, who had been there since 11:00 that morning altering the results on the cards. So we went to the computerized backup files and another unexplained event had occurred, they had all been recently accidentally destroyed! That gives you some insight into the ethics or lack of the same, in these three major hospital corporations in Detroit. When you have a patient care issue, hell you don't fix it, you just destroy the evidence and bump off the witnesses.

So poor Marple went home for a week, rapidly became confused, cold and shocky and then was admitted back to hospital in terrible shape because she had been off her antibiotics for a week. But this time they could say that she had picked up the infection at home, so it didn't look quite so bad on the hospital's records. The one's that they were so busy altering after they destroyed the computer discs, which told it all! Poor Marple went back and forth like a yo-yo after that, from the cardiac ward, to the intensive care, to the floor and back again. They put her in fluid overload again by giving her two units of blood off "die-alysis", which landed her back in the ICU, then overdosed her on the potent heart medicine Digoxin, because both Drant Cowman and his resident ordered the same thing and no one caught it, so she went back to the CCU. Then her infection got so bad because they were plating her wound cultures in the wrong media and weren't identifying what was really killing her. So they had been treating her empirically, at one point she was on eleven different antibiotics at the same time and it still couldn't eradicate the strain of super infection that had colonized in her.

She developed a clot in her heart, secondary to an irregular heart rhythm and the infection, so they put her on the blood thinner heparin and she promptly began to bleed after every "die-alysis" treatment because the endotoxins from the water would potentiate the heparin's effect. They wanted to take her back to surgery to clean out the infection in her legs and her chest, but they couldn't do it now, because she would bleed to death! By this point, Marple was confused, withdrawn and terrified, she begged her family to get her out of there. But the Hinai hospital lawyers intervened, the freaking mossad muppet piano players refused to allow them to take her to another hospital that might have saved her life, because they were so afraid for the rest of the community to see, what the Detroit News had already told them, that Hinai was a Holocaust waiting to happen and poor Marple was it's most recent victim.

So Marple cried and begged her family who called Snoozin Whorold, who then said Marple was a paranoid schizophrenic, schizophrenic my ass, Marple knew that she was going to die if they didn't get her out of there. Snoozin to cover her own ass one more time, put her on the potent anti-psychotic drug Haldol to snow her. When Marple refused to take her blood pressure medicine before her "die-alysis" treatment, because she knew that it put her blood pressure in the toilet each treatment, Snoozin Whorold increased the dose of Haldol and snowed her some more. Just to keep her quiet because she was trying to protect herself. When she tried to get out of bed in her lethargic, weakened state, to make a run for it, they restrained her on all fours and increased the dosage again. The Marquise De Sade was nothing in comparison to these freaks, these keepers of the crypts. On Nov.30th they returned her to ICU again, she was so sedated she couldn't breath and her body stank from the green slime that drained from her chest and her legs. They couldn't take her to surgery because now she had DIC, a bleeding disorder that wouldn't allow her blood to clot. Marple was doomed; her blood cultures showed fungus on top of the pseudomonas and mutated strain of vancomycin resistant enterococcus that already festered there. When they tried to slice away the infection at the bedside, without sedating her, she went into shock and stopped breathing so they put a breathing tube in. They put the freaking breathing tube into her right lung, way down into her right lung only and day after day the chest x-ray reports said the tube was in the right lung. But nobody looked at the results and every day the cardiologists, pulmonologists, nephrologists, surgeons and infection control specialists rounded on her, listened to her chest and no one noticed a f..king thing. They had been too busy in their 3:00 AM meetings with the piano players covering up the deaths of all the other patients they had offed. Now perhaps you see why we had so little remorse about closing Hinai. We were doing the community a favor and probably saving their lives.

But it was too late for Marple, Marple was doomed ! By Dec.5, Snoozin Whorold had started to talk about making Marple a no code, a do not resuscitate. She said that was what Marple wanted. Marple who was so snowed on Haldol, Ativan and morphine that she didn't know whether she was dead or alive and Snoozin f...king Whorold said she was competent to sign a consent to end her life! The freaking Witch! She knew all along the breathing tube was only sending air to one lung. Marple's infected congested right lung that had no hope of sustaining her. That wasn't enough, the next day when Marple's heart rate went up to 148 and her blood pressure dropped to 76/39, they decided to give her the potent heart drug cardiezem intravenously, but the dose they gave her was meant for a patient who had kidney function, not a dialysis patient, her heart stopped and she promptly arrested!

So they decided to let Marple go and give the family a few days to get used to the idea. They finally pulled her breathing tube, the tube they had left in the right lung all along. Marple died, not in peace but in pieces. This once proud, feisty woman died in a rotted carcass, gasping for breath, because of their gross negligence and incompetence. The kicker was that during that whole time even though they knew all about the contaminated water issue, they never checked. Treatment after treatment and they never checked it. Then to put the icing on the cake, Snoozin Whorold to cover her ass and that of her peers falsified Marple's death certificate and said that she died of an irregular heartbeat that she had had for years. Marple had only had that irregular heart beat for a few weeks and it wasn't that which had killed her! In her own discharge summary Snoozin said she died of an untreatable hospital acquired infection, but on the death certificate she falsified the cause of death to cover everyone's ass.

Marple died at 2:00 AM on the morning of Dec.11/97, I remember the date well, it was the same day they tried to kill my brother to distract us from the freaking mess that we were uncovering at the crypts. As a matter of fact, it was the same date that I recently was asked to leave (by a relative of mossad Mikey Hurst's) a hospital where all kinds of unexplained events were occurring, because they were supposedly afraid for me and afraid for themselves. The Feds, the WHO and the EPA were chewing at the keepers of the crypts tails and the freaks didn't like it. So they decided to create a few more distractions like setting fire to the electrical panel of the hospital where I worked and almost beating to death a man whose family had protected my youngest daughter for the last three years. The f...king, freaking, psychotic keepers of the crypts were back at it again, not fixing the police, or fire or contaminated water issues in Detroit, no that would make too much sense. So instead they spent their time and resources hanging World Health Care workers from the bridge and beating 72-year-old men almost to death. Assholes!

Chapter 15
Marple's Family

Marple's family was a good news, bad news situation as well. At first they put on a united front, met together, picketed together and supported their mother's issues together, but soon it became apparent that one of the daughter's was using the issue for a cash cow of her own. The first hint that something was terribly wrong was when the probate court file that named the eldest brother and sister as executors, went missing from the circuit court house that the youngest daughter had visited the week before. Then low and behold the file was found, but this time, it named the youngest daughter as executor of the will. The youngest daughter, who promptly went out and opened up an account in her mother's name, supposedly under the guise of handling some estate issues, but no estate issues were handled. Even two years after her death Marple lay in an unmarked grave, because no one would pay for her headstone. But during this time the youngest daughter, who didn't work and whose husband worked as a security guard, started making all kinds of purchases, a big screen TV, new computers, a new Ford Taurus, all kinds of things that were far beyond their means. Her brothers and sisters watched her and soon realized that she was on the take, first from Hord through a guy named Houndsend. Then through Drant Artass Whortering at PMC and finally through Benny and the Jets from the city of Detroit . All of our work that we had prepared for the state and federal government, she copied and sold back to the keepers of the crypts, then set up my friend, so that every time he came back to help us, they ambushed him. Finally murdering him, on his sixth time back, because he knew too much, and had all the information he needed to collapse the bastards. So Taresa set him up and they murdered him and helped set up elaborate ploys to confuse and fool all of the federal groups that came after him. Jason Sweins, Scary Snalexander and Prick Stein would arrange elaborate productions, Silverstein's productions they were called. To cover up the fact, that all of the patient deaths had indeed been caused by the contaminated water from leaking sewer and water pipes that they had no intention of fixing. And to cover up the fact that they knew all about the grossly elevated levels of aluminum, mercury and lead in the water, but they weren't going to fix that either, because that would cost the car companies and the city a few bucks. They knew all along too, that none of the dialysis facilities met fire code, they were all in basements with no wheelchair egress, despite the fact that sixty percent of the patients were wheelchair bound. They knew all of this three years ago and weren't going to do a freaking thing about it. But then they murdered my friend and the Feds came after them with a vengeance. They don't take kindly to one of their own becoming a MIA (missing in action) The city's complicity in the filthy water, the police brutality and non-existent fire service, were things that we knew they were all involved in, early on when we brought some family's to meet with PMC's corporate staff to try to get them to fix the contaminated water and fire issues and they brought a whole swat team down on us, including Bonaparte himself, the bad fireman, the crooked FBI person and the crooked guy from the World Health Organization. I can remember thinking whoa, what is it that these people are all so afraid of, I'm 5ft 6inches so, it doesn't take a 6ft 5inch, 300 lb cop in riot gear to do me in, let alone several of them. They let us upstairs all right, but then wouldn't speak to us, said it was a medical problem, medical problem my ass, it was an environmental disaster and that's why all the city of Detroit flunkies were there because they knew it too! So I took my notes and took pictures of them all and we worked a little harder to try to understand the scope of what it was that they were covering up.

Marple's youngest daughter turned out to be a real piece of work too, she would sedate her own husband with the phenobarb she was supposed to take for her seizures, so she could go out and party. She partied with anybody, including the mossad cardiology Drants that did her mother in, as long as they picked up the tab, paid the bill and shared their drugs. Once Taresa met Prick Stein who ran the city of Detroit's budgets the games deteriorated rapidly into very dangerous and sick behaviours. They called her Frankenstein's monster, because she would do anything, including helping to drug, entrap and murder federal agents or anyone for the plants (that's what Despeisner called the police ants) for the fun of it. Even Despeisner, who had seen it all, was shocked by the psychotic behaviours of that pair. That was one of the primary reasons that they got me out of there because he knew they were systematically poisoning me. It was obvious, I was frightened, confused and my hair was falling out in bunches, while they laughed and partied with the same animals that had killed all the patients in the city of Detroit.

So we essentially wrote her off and went back to helping the other patients who were still trapped in the system. But it was very difficult, even firms like Sharfoos and Christinot, who were supposed to be patient oriented, turned out to be a joke. One of their principal partners turned out to be married to a physician who worked for Hord's sports medicine clinic so he was of no value, in fact he told Hords so they could deflect the issue. Winters, Silver and Warts PC were no better they were too involved with negotiating for and defending big sports players to try to protect the patients. The black firms were a write off as well they were too busy protecting the black elite, to care about the poor people who were trapped in the nightmare of american "die-alysis". So once we had forced the outpatient unit to stop using formaldehyde, and once we had closed Hinai and after my home was broken into one more time on Jan.21/99, I came home.

Tony Despeisner had told me so much about the keepers of the crypts and their mossad muppets and semi-colon puppets, that it sickened me. There was no limit to what they would do to send their messages, to make their point, to win their games. No limit to what they would do for power, their ego trips and thirty pieces of silver, so I left and came home and learned all about mirror images. That Windsor had become a little mirror image of Detroit, a not attractive mirror image of Detroit. I watched as all of the things that Tony D. told me they were going to do came true and I decided, that the only real way to stop these dickheads, was to lay the cards all out on the table. In essence everyone knew anyway, at least bits and pieces, but they were all too afraid to say something. But when you analyze who, they were afraid of, that made no sense either. Why were they afraid of a couple of democratic crackers that they could easily make into mushroom soup, why were they afraid of a couple of rabbis who had inhaled too many fumes from incense sticks and why were they afraid of Archie and Pisser and Schmancy who had never had a good hair day in their lives! It didn't make sense for so many people to live in a fear state at the expense of a few bullies. When you analyze it further, you could see that this is what they chose. They chose to be bullied and intimidated into silence, chose to keep their mouths shut, chose to be victims of the freaks. Unfortunately that is why so many of them either ended up with prostate cancer or with limp willies. So I think it's time for some people to grab their balls and be men, literally and figuratively speaking and refuse to be victims of the freaks who think, only think that they have the world by the ass!

Chapter 16
Louis' Diamonds Get Rough

The whole mess had actually brought Louis number one(Louis Freeh) and Louis number two, to this area to investigate. Their good FBI staff, had told them it was a real LouLou, literally and figuratively speaking. But what they hadn't bargained for was the fact that Taresa was on the take from the beginning and drugged them all with heroin and hallucinogen laced pork chops, so that Jason Sweins and Archie's Stein could read their logs and find out just how much the FBI really knew, so they could outwit them with all of their resources. At first the LouLou's called Taresa their diamond in the rough, not knowing that everything she knew and every document she had came from us. In turn she turned over all the information on their personal habits to Archie, Hord and Pisser's boys, in exchange for even more cash. She found out that the LouLou's had a short fuse and were quick to make assumptions, so she fed into those flaws with the freaks from the crypts help, so that the LouLou's would close their investigation and go home, so she could start enjoying some of the money that was accumulating in her name from the keepers of the crypts. When one of the Lous' passed a ring to her, to give to someone else, she wore it for a little while then turned it over to Jason Sweins for ready cash and a newer and bigger car. The Hords, Assr and the Sweins roared at this, they thought it was so funny that they had duped the FBI and were continuing to get away with the genocide of the 1,200,000 people who lived here. They didn't care, their pockets were full and they could afford complex water and air filtration systems in their own homes, so there was less of a chance that they would succumb to the environmental holocaust that was killing every one else. They didn't have to worry about where the agents were either, because Homerica was tracking them for Pisser and the boys of the "ICK Club". Ed and the two temps at ATT and Shameritech were monitoring all their phone calls, so they knew exactly what the FBI were going to do next. Any of the good corporations who tried to help them were taken care of by Michael Powell and the sham of a justice department, who went after only those who wouldn't bow to the democratic demands for protection money. Soft money donated to their campaign war chests in exchange for immunity against discrimination and anti-trust lawsuits. Those corporations who wouldn't donate willingly, like the tobacco companies, they went after tooth and nail, ravaging them in court in class action lawsuits designed to feed mega bucks into every democratic coffer in the US. Companies like Coca Cola, Texaco and now Chrysler's they set up in sham discrimination lawsuits, to teach them the lesson that if the keepers of the crypts couldn't get the corporations money willingly, they would get it anyhow, through Boss Hawg government tactics. With the Hawgs now being the cracker democrats, the investment banks, the bad Jews, the Hords, the Jackass Coalitions and the rest of the bottom feeders that fed on the carrion of those they killed, literally and figuratively speaking. But the LouLou's do know the truth now, because we told them and brought them and the federal justice department the proof as to who the creeps were going to off next and why. So this time the only ring Sweins and Stein are going to have is one through their nose, as they are led to the federal pen, for their complicity in multiple murders. What a fitting end for the freaks who tried to fool the Lou's. They have met their own Waterloo, because they couldn't keep their big mouths shut. Tadum!

Chapter 17
The Ewegenes, Good and Bad

The first Ewegene that I ever knew was Ewegene Applebum, who was a friend of a person that I knew who worked for the Hord crypt. He was kind of funny at first, he followed me around, even when I went on vacation in Europe. He was a pharmacist, whose hobby was photography. When I would take my films to be developed at his stores, I would get all kinds of beautiful pictures back, double, triple and quadruple exposures of the cathedrals in England and France, arranged in collages that looked like true works of art. I knew that I hadn't taken them with my little disposable cameras and for a long time I didn't know who did, I just thought it was funny and sweet. That was until the Hinai disaster happened and I took a roll of film that had pictures on it of the goons in the boat, outside my home. I took that film to be developed at his store too. The film had 36 pictures on it and I asked them to enlarge the last three of the goons in the boat to 8x10's. When I went back to the store to pick up the pictures, the clerk was very nervous and said there were no pictures of the guys in a boat. She was standing there with a roll of film under her hand, when I asked her if that was mine and she started stammering, I just snatched it out of her hand and took off. I took the film down to one of Raubman's stores and had it developed again. Low and behold Harbor's had only given me back 30 pictures, Raubman had given me back 36. The only pictures that Harbour's had kept, were the three of the guys in the boat and three of me. I no longer thought it was sweet or funny if they were protecting the bad guys who bumped off our patients and at first I couldn't understand why, until I found out that it was illegal for board members to have contractual relationships with the not for profit boards on which they were served. Which was a real big joke, because all of them had. The boards of the crypts were really southern in their relationships; that is their relationships were incestuous to the nth degree. So I lost a lot of faith in that Ewegene, his little jokes weren't funny anymore. But what is funny is that he is now on the board of an even bigger company that is doing the same darn things. I have kept the pictures though, the collages are pretty and remind me of the way he was before, he was stained by the crypts.

Then we inherited Ewegene Schiller, what a piece of work that one was, he would make a deal with the devil just to make a buck. He was in charge of "banking services"' for the crypts and the mossad muppet club which meant he got to bribe the heads of other banks, Canadian banks, to keep the crypts intact. He was also in charge of tracking federal agents by keeping track of their withdrawals at ATM's. Talk about, full service banking! That Ewegene would do anything to keep the deposits coming in. Sham company deposits, didn't matter, shell company deposits didn't matter, Homerica brought money laundering to a new level, the board of trustee level. Homerica knew about the water problems all along, Heinz Prickly and several others including some who were involved in the Mercury Abatement Project in the New England states sat on Homerica's board. They all knew but chose to do nothing, it was bad for "bidness" they said, cost a few dollars they said, and when that Ewegene showed the slightest bit of conscience, Pisser did his pimp routine and married him off, so he could go grocery Grovering to keep him occupied. But God did a funny there too, because that Ewegene goes to work every day at the top of one of the tallest buildings in Detroit and that Ewegene looks out his window over at the city of Detroit building and knows that if Homerica has a fire, that he is done like a dinner. Roasted his own chestnuts on an open fire so to speak, because Archie and the semi-colon puppet club was too busy hosing him, to buy hoses for the city of Detroit fire trucks. There is a very important part I forgot to add under the Homerica Ewegene. The riddle is, "When is a recession, not a recession". Answer: When it is created by democratic big "Bidness" and their international investment bank cohorts who strip out the revenues of large companies and hide them in willing banks, before a new government comes in. To cripple that government and make it look like it was the cause of the downfall of the economy. A lack of confidence vote in the new leadership. A cheap shot by Billy and the boys to empty the U.S. coffers and run one last time. And they did, they cleaned out all of them, Chrysler's, GM, even Hords, who was so busy chasing us that they didn't realize that they were being screwed by more than one person without a kiss . Then the Jewish "ickonomists, true to form trumpeted the possible casual factors of the recession when they knew all along who had done it and why.

A new lesson in mossad and semi-colon "ickonomics", Pisser and Schmancy and Cracker democrat "ickonomics" . Barclay's, Fidelity Magellan and First Suisse Boston bank economics . And here all this time you thought I was just napping. Guess what, you were wrong!

There was a third Eugene, but that issue was not funny. The third Eugene was a truly good man who was married to someone who worked for a local channel seven new's station. A news station who knew Archie's bunker for what it was and the keepers of the crypts for who they were. So they set up a sting operation to capture some of Benny's goons, the same bad cops who did in one of their own at the dry cleaners, because she had exposed them too. They arranged an accident that was not an accident, filled his inhaler with atropine instead of ventolin and sent him on a hayride where he promptly died of a respiratory arrest. They did it just to send a message that no one f…ks with their "god", their goons of Dennis. But guess what, the justice department knows that now too and they know that Benny will never find the murderer of the police officer that they hung out to dry in the dry cleaners, because she had hung some of bad police out to dry. John Ashcroft knows that Benny will never find the murderer because they did it themselves, just like they did in all the rest of the bodies that are mothballed in the HP plant. No one messes with George Herbert Walker Bush and John Ashcroft and gets away with it, not Berry not Archie, not anyone. Archie and Benny made the wrong choice this time, when they and their democratic cracker freaks conspired to set up 900 FBI officers in Ashcroft's own backyard they goofed. No one poops in his backyard and gets away with it. So like berries on the vine, a hanging vine, this gets sweeter all the time. No wonder Tony D. made like a disappearing wabbit and wan, he needs to, for all the mossad and big "bidness" murders he has been complicit in he needs to disappear, forever!

Chapter 18
The Ho, Ho, Ho's of Homerica!

Ewegene wasn't the only carp in the Homerica pile, the board of trustees read like a southern cousins reunion. First and foremost was asshole windbag Snale Horden, the CEO of the Hord Crypt, who not content with the honour of having done in Fendry, was now going for the gusto in trying to off half the city of Detroit. He knew all along how grossly contaminated the water was and that the company he worked for and three quarters of the people on Homerica's board were many of the key reasons why the river was a toxic sludge dump. That's why they hired David Barker Schmoois of Schmoois and Stunday, who was also on the board to cover their collective asses. The pimp Pisser was a corporate director as well, in charge of pimping 101 for their operations, with his boy Peter Goings. Heinz Prickley too, who owned the black lagoon filled with PCB's and petroleum products and who peddled misinformation to the national security agency. Then we had Hen Wrongway CEO of the Schmer Corporation and Spiergallani head of Novartass. The cast of characters read like the screenplay from "Cats" on a Hot Tin Roof. A whole platoon of mossad muppets and semi-colon puppets, who were all tap dancing as fast as they could in hope that no-one would see what was so very obvious, that the people of Windsor and Detroit were dying like flies and the trail of pestilence led directly back to them. Talk about "Deliverance", all they needed were the banjo players and they brought them in too, in the form of Al Gore and the Democrats, to protect their collective, "honour", which in this case should have been spelled "horrors". What a sideshow, what a freak show, I would love to have the unabridged meeting minutes from that 31st floor boardroom. The Ho's of Homerica, hose two countries. Wouldn't you love to hear Snale Horden's arguments for injecting people with formaldehyde every dialysis treatment. Hell who needs Herman Goebels when you've got him. He and his cronies gave new meaning to the term holocaust, they should have called it homericacaust. No wonder they had to hire Guard Farce to keep their secrets secure. They were systematically offing more people than the third Reich and they wanted to continue getting away with it because it made them and the investment banks that owned them a few bucks. The Homerica board was a brothel of three-dollar hookers and Maxie, good old mossad Max Pisser who gave it to everybody in the ass, was their pimp!

And of course Maxie not being content with decimating the populations of the cities of Detroit and Windsor, decided to go after Chrysler's too! Easy to do when you had Snale Horden on the board, CEO of Hord's health system. They just got Chrysler's legal council who worked for David Barker Schmoois who was also on the board to facilitate stripping out the revenue with Schmeaton and Ritchie's help. This made the Schmer corporation and Prickley happy too, because they all held stock in Hord's. Easy to launder large sums of money stolen from Chrysler's when Homerica is more than happy to help. Eliminates Hord's competition, at the same time that it strengthens Maxie and the mossad boys money bank. Then they can just concentrate on decimating GM, which they had already started with Harry Pieces help by planting a Hord Chief Financial officer in the GM camp, to bleed out GM by paying all the profits out to sham companies. Their sham companies Maxie's, mossad's, the Ewegenes, Semi-colon's and Hord sham companies. Easy to do when you have CEO's like the Sunbeam boys who are so focused on their own bonus check, that they can't see the forest for the trees, let alone see mossad massacring them. Amazing isn't it that any company can survive the US Gulag Archipelago, but rather simple to do, once you know the games the boys of the crypts play. Then you can let the dogs out after them, and play, "Trick or Treat" with them, or "Tag Your It", with them and drop their stock 63%, just like we did to Homerica, or 49% and dropping, just like we did to Hord. Easy to do, when you know all the tricks in their cookie bag. Easy like Sunday morning!

Chapter 19
Heinz Pricklies Baby Food

When people made references to me about Heinz Prickly, I didn't have a clue as to who they were speaking of. The only thing I knew, was about the time they started referencing him, Stein's animals and the bad Drants started poisoning my food, putting sleeping powder and amnesiacs in the food I ate at work, so I would go home and sleep all day and forget about pursuing the patient care issues. They were aided and abetted in this by a young psychiatrist named Piccinini, who truly needed to see a psychiatrist himself. He was Hord's expert at the clinical dosing of psychotropic drugs so that they would achieve the desired effect. Making people functional when you needed them to be, but then keeping them down when necessary . They were desperate to keep me down when the Feds were in Detroit. It wouldn't look very good if I was picketing outside while the decoy that they had pretending to be me was meeting with the regulatory agencies inside the building!

My god the extent these people went to, to cover up the mess, no wonder the EPA fined Detroit 200 million dollars for dumping raw sewage in the river; they were so disgusted with the freaks from the crypts that they were beside themselves. At first the Drants thought Heinz was out to protect us, but in actual fact what he was interested in was protecting his backside too! Heinz was very nervous about two things, number one the Black Lagoon, the downriver cesspool so full of PCB's, petroleum and toxic waste had his name written all over it. The water was so grossly contaminated it was like black jello. Number two, Heinz whose ABS company supplied information to the National Security Agency and several other federal security agencies, had been fibbing to those agencies all along. Heinz sat on the board of Homerica, with the Hords, the Pissers and the Bunkers. But they didn't give a damn, switching to an ignition switch that didn't use mercury would have cost them a few cents a car and would have eroded the bottom line and they couldn't have that.

So when Heinz showed signs of having a conscience and doing the right thing, the Hord Drants started to drug me to keep me down and mossad Max Pisser did his pimp routine, setting Wally up with a mossad muppet named David, to channel Heinz' dollars into the jewish coffers in other words Pisser's coffers too! Pisser to protect his scams still has his freaks following me around, just ask Steve Defalco the new P.R. guy at the Windsor hospital where I used to work. That's why Defalco was given the job, to protect the jewish honour! Jewish honour my ass. Tell Lady Di's sons and Trudeau's surviving sons about your freaking jewish honour, I'm sure they would love to know! Assholes.

And while you are at it, you might ask Defalco, how in the hell, did "jakobs" disease, "mad cow" disease, get into a Windsor hospital? Another intimidation tactic to keep the Drants quiet about the people that are dying like flies from cancer, birth defects and heart disease in Windsor? Or perhaps a little message from the WHO, who were pissed that Windsor helped cover up the hanging of the World Health Care workers from the bridge. I bet Defalco drives a Hord product, just like the CONsultant who manages Windsor's water supply! Amazing isn't it, but you have to admit it takes a lot of effort to cover up the slaughter of 1,200,000 people. Just ask Hord and Archie, it's costing them a fortune in bribes, almost as much as it would have cost them to fix the problem three years ago. But I guess the whores at Fidelity Magellan, Vanguard Index and College Equities Fund, who own Hord, don't give a damn about that. They were too busy buying AOL Time Warner and selling the US secrets off to the Chinese with Powell, Rice and Clinton's help, to worry about our children that they are murdering here through environmental genocide. Bummer isn't it, now they will have to continue the cover up and fix the problem too! My, my, my, couldn't happen to a more appropriate group of corporate pimps, who sold the welfare of our children and their families for thirty pieces of silver! Assholes!

Chapter 20
Health Care Ickonomics, Detroit Style

I have a simple riddle for you. "How does a health care "bidness" make its money?" By making people sick silly and keeping them sick, at least those with good third party independent insurance. Do you think HMO's like CRAP, made $149,000,000 in profit last year simply by pickling dialysis patients alone. No way Jose, they diversified, they knew about all the Canadians that they were killing off with environmentally caused cancers. They knew for two reasons, number one because they had caused them and number two because they were paying the oncologists, pediatricians and cardiologists mega bucks to keep the real statistics quiet. The keepers of the crypts knew that there were over one hundred cases of cancer diagnosed each month, in a tiny city like Windsor. One hundred new cases that if the freaks, played their cards right, they could milk each victim, or that victim's government for $250,000 for each case. Easy to do with multiple trials of chemotherapy, radiation and surgery. A guaranteed new cash cow for the crypts. Twenty five million a month, three hundred million a year, if they did it right and captured all the victims of the cancers that they had caused. Talk about vertical integration, the crypts could birth them, kill them and cure them and keep all the revenue generation in their food chain. All they needed to add was funeral home services and I'm sure they will do that now too, if there is a buck to be made in it. Do you see now why the Billy Hay Hords and the Drant flunkie from Sierra Leone have no interest in cleaning up the environment. Hell, if they spent a few dollars on toxic waste containment, it would cut into the revenue generated by the people that they were killing off with cancers. Billy would have to buy one less pair of flatties for his fanny and one less flag to fly over the corpses of the birds and fish that are being killed by the same toxins that were doing us all in. One less pair of flatties, now wouldn't that be a bummer!

I have a new ad campaign for Assr and Swines, they can do a panoramic picture of the five hundred and sixty three children who have been infected with cancer in the city of Windsor. In the picture they can have Hord pennants in bribery blue, waving proudly on flagpoles over their heads. The caption can read, "Think Hord First, Especially on Autopsy", as one by one each child is faded from view, symbolizing their lives that Hord had stolen from them. Next they can do a collage of all the grandparents of those same children, slowly fading from view, one by one. Victims of the cancers, heart disease, and Alzheimer's that have been killed by the environmental genocide. Then overhead we can have flags flying, but this time with a yellow Star of David sitting on that field of bribery blue, symbolizing Max Pissers hand in the environmental holocaust. Mossad Max Pisser and his crones who pissed on the Lincoln Memorial, because they were too busy collecting their thirty pieces of silver, to see the carnage around them. Finally, we will add a black border to the flag to symbolize the black elite, who knew all along about the cause of the genocide and sat on their hands and did nothing. A perfect pennant for the evildoer's team, the triology of thirteen team . A bribery blue flag with a yellow Star of David, bordered in black, flying over all of those that they have been victims of the new holocaust . The one that still continues today because Maxie holds the keys to the ovens this time and if there is a way to make a buck he'll do it. Maxie, who parlayed the innate hatred of the Jews against the Germans, the Palestinians, the Catholics anyone as long as it made him a buck, into a personal fortune . The Max Pisser factor, who brought new meaning to the "Think Hord First at the Mausoleum" motto; who used his ngrants as he called them to do the dirty work, so as not to stain his shrivelled yellow hands, that gave the claws of Macbeth new meaning. Maxie whose favourite colour is mustard yellow, the colour of liquid shit that eddies and flows around him wherever he goes. Maxie who is responsible and accountable along with the freaks on his team of evil doers for the systematic genocide of over 1,200,000 people. We can put mossad Maxie in the picture too, waving his flag in the middle of the now empty brown Hord field, with 1,200,000 crosses in rows around him. Waving his flag and counting his change, the schill with his thirty pieces of silver!

May 2001 addendum: Three weddings and a funeral. A classic exercise in jewish greed, some mossad morons removed the support pillars from a jewish wedding hall to cram more people in, to make more money. The result; twenty five deaths and hundreds of people being injured when the building collapsed. All because a few schills wanted to make a few more bucks. Three weddings and lots of funerals, what a way to celebrate, what a legacy to leave! Assholes!

Chapter 21
The Pikorny Club Boys

How do you keep a bunch of surgery boys occupied, you give them a bunch of free whores and water piks and let them go at it! The Jewish mossad mafia was a pretty clever bunch. They knew the best way to keep a bunch of bored and frustrated doctors occupied was to provide them with a bunch of free whores who would suck them off between cases and provide a little night time entertainment whenever they decided to have a late case or an "emergency". Dr. Piece was the one who orchestrated that exercise along with his friend Dr.Cowman from the US., who knew that they needed to keep the boys occupied, who were helping to unearth what a sham the care delivery in the Detroit crypts was. Both Dr. Piece and Dr. Cowman knew all about the fact that they drugged and harassed nurses in the city of Detroit and Windsor too, for that matter. They tried to destroy anyone who had the nerve to question their incompetence, no matter how blatant it might be, they fancied themselves as untouchable's, sexual adventurers who could get away with anything because uncle Maxie was covering them with his umbrella.

It's funny, in any city, in any country, good physicians, competent, caring physicians are never threatened by scrutiny. As a matter of fact, they scrutinize themselves, because they truly do care about their patients and their outcomes. If there is a way to do it better, then they do, because they understand that someday it might be them or one of their loved one's under their peers blade, so they want to make their teams the best of the best. John Hopkins understands this, Beth Israel New York understands this, even going as far as allowing their morbidity and morality rounds to be televised because they want people to know they are truly trying to be the best of the best for their patient's benefit. Hord's cardiovascular team used to understand this, back when Don Magilligan, Dr. Keith and Dr. Oh practiced medicine there. They were men who would stay up all night and pace the rooms after a rocky case, a high-risk case that no one else would touch. They would do the surgeries as that patient and families last hope. Then stay with us the difficult first hours of recovery because they wanted to make sure the person who came to them with their last hope, the person who trusted them with their life, the person who laid in the bed, was going to be okay. That is why I have such utter disdain for the Pikorny club boys, who rely on the piano players to cover all their mistakes. Once you have had the privilege of working with the best, the rest all look like Clintons. Some of them might say there is no place for excellence in a small town hospital, but that is bullshit. Even tiny hospital's like the Veteran's Administration Hospital in Sepulveda California, who truly care about their patients spend their tiny bit of resources in the right place, because they place their patients first and foremost. They know that with excellent patient care, the financial deliverables will follow, as their patients stay well and healthy.

But the crypts, my god, the crypts that own their own doctors, their own hospitals, their own HMO's and their own off-shore, for profit insurance companies, they place their short term profit before everything and everyone. The crypts that will do anything, including setting fire to small town hospital's electrical panels in order to bully and intimidate them into silence. The crypts that make $149,000,000 profit per annum from their HMO constituents by increasing their premiums 14% per annum, while delivering third world care. They will tell you themselves clearly and directly, the cheapest patient is a dead patient. So the HMO's collect the premiums, until the patient starts to consume resources, then they off them early by delivering minimal care so they don't break the bank by using up some of the resources that they and their respective companies have spent their lives paying for. The sad part was, that they got away with it, used to get away with it, until the people who were paying the bills started asking questions about where their money was going and how come a not for profit health care system was able to rape $149,000,000 from it's members in a year, while the morbidity and mortality rate for the area was thirty percent higher than the rest of the country. Now these same people are asking questions, key questions, pointed questions about the care delivery of the crypts and as a result are taking their "bidness", somewhere else. Whoever said there wasn't a God, was wrong, sometimes he/she just has to wipe the sleep out of his/her eyes! Then perhaps while he/she is doing that, God can help the Pikorny club boys get their heads out of their asses as well!

Feb.28/01 postscript, a sad thing happened this week, a travesty, a nursing school classmate of mine was murdered by three of the incompetent freaks of the Pikorny club. A young and beautiful woman, she entered the hospital last Thursday with a gastric ulcer, caused by two years of illness, from a simple knee injury that turned into a black comedy of medical treatment. She came in on Thursday and was dead by Sunday, from a simple gastric ulcer that the boys from the Pikorny club Hackus, Quackpata and Cogless, failed to treat promptly and appropriately. Cogless the surgeon, who almost killed my brother through his incompetence two years ago, has struck again, but this time their handiwork was fatal. Chalk one up for the college of Caribbean and the Cogless clan, a beautiful young woman is dead, because the crypts big and small, on both sides of the border protect it all. There is no system of checks and balances not with the bunker puppets and jewish mossad muppets running the show. But now maybe Nancy can sit in the car beside the boys from the Pikorny club wherever they go and whenever they turn their head, they will get a glimpse of her beautiful face, with tears rolling down her cheeks, because she will never see her children again. I hope that Nancy will haunt them forever, wherever they go, what ever they do, so that they never ever murder someone like her through their gross incompetence again. Maybe God in his/her infinite wisdom will see to it that they die young as well, on the cold slab of an operating room table, at the hands of one of their equally incompetent peers. That would be the most fitting retribution, for all the travesties, they have committed. Animals, like mad cows and english men come in all shapes and sizes, some even come in O.R. greens.

Chapter 22
The Belalagosy of Blow Job Bill

The Blow Job library is supposed to store all of the archives of his presidential legacy. That should be a fascinating read, where does one start. How about, in 1985 when as governor of Arkansas, he single handedly sentenced the majority of the haemophiliac populations in Canada and the U.S., to certain death by selling unscreened HIV contaminated blood products from the Arkansas prison population to both country's Red Cross organizations, in order to make a few bucks for his state government's campaign coffers. Or shall we start with his personal, "friends", who knew too much about him, who ended up in prison or as "clintoncides", supposed suicides, to shut them up! Do you know how many former friends of the Clinton's have ended up as suicides? Enough to field the Lion's football team, amazing isn't it. Of course the library will have to include the lip prints of all the women that have had to blow him in order to keep their jobs over the last several years. Then there will be the cigar store section with a selection of flavoured cigars, the ones that were his moral alternative to adultery. He wasn't having sex, if he flavoured his cigars in unique and creative ways, just satisfying his oral fixation and who could blame him, Billary wasn't exactly Blow Job's answer to Norma Jean. In the national security advisory section, we will have all the archived records of Blow Job being advised to wag the dog and go to war with Iraq, to distract the nation when he was found with his cigar up Miz Lewinsky's patitouche. As well as the security advisors recommendations not to blow the cover of the FBI agent the mortician, until the Republicans were in power, so it would look less embarrassing for Billy. In the white house china section, they can list all the secrets that Blow Job and Billary sold to China and Israel and Russia too! Just like in a china shop, they can put price tags on each piece, to show how much the C's squared, sold their country for to put money into their own campaign war chests. In the white house menus section, they can list mousse, spelled Mossad, the Israeli spy organization under their favourite dessert, for how richly the Israelis' sweetened their pot, so that Billy would fight and win the holy war for them and pardon their felons at the same time. Under trade and communication, they can list all of the companies like AOL, Oracle and AT&T who they were in bed with, who provided information in exchange for political protection against Powell's antitrust witch hunts designed to make them wealthy men personally and democratically. Some of the dems even had an insurance policy built in, because by acting as mules, leaking tidbits of information between the two parties, it ensured them a political parachute to the Bush camp, in the event that Blow Job lost his Teflon coating. In the defence arena, we will have all the documents outlining who needs to defend who, against whom. Telling the details of how the Bosnians needed to defend themselves against not their enemies, but Semi-colon and Blow Job who allowed their troops to use radiation contaminated munitions, which will decimate their country for years to come. And the fact that Al Gore and Arizona's McCain were going to have to wear lead lined underwear to bed, if they won the election, so that Billy's Israeli friends could have one of their own placed in power by default. Easy to have a mossad led country, just ask Maxie Pisser, all you have to do is get the "right" people elected, then bump off the non-Jewish competition, easier than pimping Billy Hay Hord and Heinz Prickly off to a fanny, and much more lucrative. That way you could control the "Ickonomics" of the whole country and terrorize the whole world with "non-violent crisis intervention." In the closet crisis intervention, where your semi-colon puppets pray for peace publicly and preach the parables of diversity publicly, while they murder everyone privately in accidents that are not accidents and in allied frames, just like when they murdered Louis Freeh's friends, the Giovanni's in Italy. The games they play are meant to dupe the people with short fuses, who can only grasp the seemingly obvious. In order to be successful in dealing with them, you have to do your own research to understand who each of the players are, what their agenda is, who is blackmailing whom and for what and most importantly you have to know the keys to their crypts, all their crypts. So far we've only let you into the parlour, but fascinating isn't it, amazing the little bit that we have shared with you. Intriguing to know what will come next and why and if you are one of the good guys we will share it with you, so that you will know how to pour draino down the ant hills of the evil doers, and save the world. Just like we saved the people of Detroit from certain death at Hinai and just like we are trying to save you right now from an early death from the environmental holocaust in this area that is killing us all. We have the keys to all their crypts, even Blow Job and Billary's crypts too and if you are very, very good, soon you will know too!

Chapter 23
The Dog Gets Wagged One More Time!

It truly amazes me with the U.S. having the Central Intelligence Agency, the National Security Agency, the FBI, the Armed Forces Intelligence Services and myriad other security agencies, that they don't know what is going on in their own country. It also amazes me that on the advice of a few people who are trying very hard to hide the skeletons in their own closet, the whole country goes to war repeatedly. You would think, once bitten, twice shy, however this doesn't seem to pertain to Billy Boy Clinton and his ruses. So it would appear that Billy Boy and his closet cabinet that now work for Bush, have dusted off an old strategy to deflect from the critical issues as to why he pardoned a prisoner with Israeli spy linkages and why he set up 900 FBI officers, by threatening to pardon Leon Pelletier. Now very clearly one begins to wonder where the heck were all these people, Deutch, Powell and Rice, who were supposed to be protecting their national security. Where were they when Billy was giving away the store to the Chinese and Israeli spies? Certainly not doing their job, were they hiding under the bed, or perhaps in Monica's closet or maybe, just maybe they were complicit in the interactions too? And is that why they are so busy trying to get Bush to Wag the Dog and create a distraction to deflect away from a full-scale investigation of Clinton and his cronies selling military intelligence and pardons like a sheets, no sheets motel. I guess that they figure that since it worked the last time, to take the edge off the Lewinsky thing, that they might as well resurrect the ruse again and get as much mileage out of it as possible. Maybe if they escalate the war rapidly enough, they might be able to get the FBI, the Congress and the house to drop the issues altogether, since they will be too busy with other things. Too busy to look at exactly who was involved in the selling of information, who was involved in setting up the FBI, who was involved in the planning of the multiple accidents that were not accidents that took place during Clinton's reign. Not very attractive when you bump off the competition through plane accidents, car accidents and facilitated heart attacks, not very attractive at all!

May 2001 addendum, ho, ho, ho, look who is in Hong Kong selling secrets once again to the Chinese, after being, "briefed", by Powell and Rice. Surprise, surprise, it's Billy Boy and look who's footing the bill, AOL Time Warner, owned by the same freaks that own Hords. Amazing isn't it, a freaking southern cousins reunion one more time, willing to sell anything and everyone, even their own country for a buck. What a joke, the United States of Mossad, Bejing, Moscow, sold off by their "finest", for a free lunch and a little loose change! Perhaps they should teach that in American History 101, it would be a real eye opener for the boys of the CIA, NSA and FBI, but then again maybe if they taught it at Quantico, it wouldn't be so hard for the boys to figure it out!

Chapter 24
The Horse's Ass Attorney Generals From the Democratic Crypts

Most people don't realize that 38 of the U.S. states have democratic attorney generals. The sad dynamic is that these 38 fools are so bent on guarding their own political agendas that they are destroying their country and the people that they are supposed to protect. They have no qualms about decimating their country's economy, as long as the spoils go into their own war chests. No qualms about turning a blind eye to environmental holocausts, if it means keeping their re-election cookies in their own bag. You show them the data of lead and arsenic contamination, five times the safe limit and they see nothing, you show them the mercury and aluminum levels at one hundred and fifty times the safe limit and they hear nothing. You show them death rates 30% higher than the rest of the country and they are mute. They are "pants," political ants who protect the carnage from the bad corporations, merely so that they will have a few more cheap lunches and fund raisers thrown their way. Want them to address the Brazils of Detroit and Washington, forget it, they are too busy issuing court orders to illegally tap phones to protect big "bidness". Want them to address the cops that are murdering so many people in Detroit, that they lost a senate seat, because the population dropped so much. No way because "Hizonner" is a democrat . Want the attorney general to intervene in the horrors of Hell Michigan. Hell no, they are too busy busting kids with a reefer in their pocket, to find out where the $62,000,000.00 in cocaine that had disappeared from Benny Napoleon's police store room has gone. Too busy at fund raising breakfasts to protect the firemen and news reporters that the boys from the bunker are trying to kill for exposing their charade of city "gubermint". And when the attorney general's office is approached about the attempted murder of World Health Care workers, they, like Benny and the Jets, know "nuttin" and the other attorney general's office too, at least the democratic attorney general's offices, 38 of them, parrot the same refrain. "We don't know nuttin". As long as Hords, Homerica and the big investment banks are sponsoring fund raising luncheons, you can continue to buy piano playing public defenders for less than a big mac! Not very encouraging is it!

Then you have the activities of the Democratic attorney general's enmasse. Do they focus on improving the welfare and well-being of their constituents who are being offed like flies by third world health care? Hell no, they focus on busting companies like Microsoft, to put more dollars in their democratic NAG bag, to ensure their further election, so they can continue to ruin their country through partisan politics and trough warfare. Do they support the WHO, the EPA and other agencies, who are trying to stop the environmental holocaust ? Hell no, because it might alienate the, "big bidnesses", that bought their positions for them in the first place. Is there any hope that these dimwits will start to make the "Right Choices" and stop selling their country like jamaican tee-shirts? Of course not, especially with Billary, putting their PAC Team together. Once that schill game gets going, the democrats might as well buy a kiosk on Lundy's Lane, because everything and everyone will be sold for a vote, like cheap plastic hord keychains at the local five and dime. And once mossad muppet Carl Bevin gets control of the defence department, you will be able to buy US secrets and technology, more cheaply on the streets of Tel Aviv or Shanghai than a cup of Timmy's coffee. Such is the legacy of the democrats, most of whom you can buy for less than a three-dollar hooker, some even for loose change. So much for honour, or democratic Hiz and Her Honours . Thirty-eight democratic attorney generals; fools who sell the lives of their children and their loved ones for a free lunch . Assholes!!!

Chapter 25
The Raider, Laidlow, Fed-Icks and Hovistar Sucking Companies!

What do you call large evil corporations that will sacrifice the welfare and well being of 1,200,000 people and all of their children? The answer, corporate whores to the fourth power, who are so bent on destroying their competition, that they would collude in covering up the environmental genocide that is being committed here, in order to protect their insider trading deals with the freaks from the democratic crypts. Fed-Icks, with home bases in Billy and Al Gore's home states of Arkansas and Tennessee are more than happy to transport confidential documents and tapes for you, the only kicker being, that they will transport them directly into the hands of the very creeps that are creating the holocaust, to cover their collective asses. Laidlow, who transports children in unsafe Hord school buses, thinks nothing of huddling in the corner with Hord, to cover up all the deaths related to both bus accidents and the environmental carnage that Hords' has caused through their environmental holocaust. Particularly because Laidlow also owns funeral homes and all the deaths in this area are generating big bucks for them. Hovistar is Assr's solution to ensuring delivery to his crypts, but do you really think any truck driver would want to transport goods to a company that has murdered ten workers and injured twenty-two others in one year, at one plant alone. Assr must think everyone is as stupid as he, Troutman and the Hay Hords are. Do you really think anyone wants to have anything to do with a corporation of serial killers, that continues to build monuments to bowel movements like the Hord field, while their workers and the public are being slaughtered enmasse through their negligence? I mean really, anyone with an IQ of over five points can clearly see the freaks from the crypts for what they are, simply by their actions. One picture is worth a thousand words and the multiple pictures of all the dead from plant accidents and hexplorer rollovers speak volumes, as to where the keepers of the crypts and their partners true priorities lay. Raider was another joke in the pile, raider whose job in transporting the Florida ballots, was to lose a few along the way to ensure that Al Gore, to whom they had already paid protection money, would win. Raider didn't want the Jesse Jackass coalition to come and beat up on them through sham civil rights infractions, so they made their "voluntary" donations to the democratic coffers and were hoping to cash in on their largesse when Al took office. But Al screwed them over too, by providing insider information to Fed-Icks, so that they could take over a large part of the lucrative US trucking business. That was a you scratch my back, I'll scratch yours deal, which ensured that in exchange for the lucrative U.S. trucking business, no information harmful to the democrats would ever get delivered. This sham was almost as bad as the US postal service, where you could bribe any postmaster to intercept mail for a few bucks and a cheap lease on a Hord too. Just check out how many postmasters drive Hords and you will see exactly what I mean. Amazing isn't it, it is deliverance complete with the banjo players from Arkansas and Tennessee, able to control the mail delivery to the entire US, including the FBI, CIA, Secret Service and NSA. Heck who needs to worry about the Russians, Israelis and Chinese sabotaging the US, when it is already on an internal path to self destruction. You don't even have to buy a ticket, you can just sit back and watch and it would all be very funny except for the fact that in their struggle to off each other, they are trying to take us with them too!

Chapter 26
Assr the Nitwit

I had a hard time trying to decide whether to call this chapter Assr the Nitwit or Assr the Mossad Muppet. I think both are highly appropriate. Assr was a classic study in stupidity. He believed not in fixing problems, but in covering them up with the help of his spin master Swines. He also believed in taking risks, risking other people's lives, by putting defective products on the road and hoping like hell they wouldn't be linked to him. He also believed in bullying, harassing and intimidating people into silence and conducting pogroms of his own. Setting up sham evaluation processes, so he could get rid of anyone that opposed his little Napoleon tactics. Assr was in this mess up to his eyeballs, because he knew exactly where all the mercury and aluminum contamination was coming from, from the plants that he refused to approve expenditures on to fix critical issues like the Rouge boilers that were ready to blow or the discharging of toxic waste into the river. Little Assr was truly a man who thought only about today's bottom line and in doing so systematically destroyed any credibility and integrity that the company might have had left from the years when Sweterson was CEO. Hard to have any respect for someone who blocked the state from inspecting the Rouge plant after the explosions, so they wouldn't find what the workers knew all along, that it was a disaster waiting to happen. Hard to respect a man who sent his own wife on safari with a federal court judge to compromise him and blackmail him into compliance. Hard to respect a man who sat on his hands and watched while family after family was devastated in accidents caused by faulty suspensions and fractured steering columns on their trucks. Sat on his hands and said there were no problems, as the death tolls mounted into the hundreds. Hard to respect a man who clearly knows why the death rate in this region from cancer, heart disease and birth defects is thirty percent higher than the rest of the country. Knows it so well that he keeps local oncologists, pediatricians and cardiologists in Hord's pocket to ensure that the death certificates and the statistics don't accurately reflect what is going on here. Hard to respect a man who subsidized the salaries of three new Windsor emergency room physicians, just so they could keep an eye on us and belch to Assr, Hay Hord and Hen Hord, what exactly we were uncovering on this side of the border. The same physicians they pay not to see injuries or do x-rays on anyone who HORD has tried to bump off! And it was amazing, that as soon as we started tracking the morbidity and mortality rates in both children and adults from environmental causes, it was absolutely amazing how unexplained events started to occur here too. Like World Health care workers hanging from the bridge and mould infestations in the Windsor hospital's new construction and fires in the hospital's electrical panel. It was absolutely amazing how soon after I showed the semi-colon puppet director of perversity at that same hospital the pictures of all the people who had covered up the patients' deaths from contaminated water in Detroit, pictures of people she knew, because she was involved in the cover up too, absolutely amazing how quickly I was asked to leave, before I showed the pictures to other people too! Her job was to protect the sintegrity of the black community, who had also known all along about the high death rates and the contaminated water and had not done a freaking thing about it. Then to put the icing on the cake, the same fool, recently solicited funds, $200,000.00 not to protect the welfare and well being of the children in the community, but to erect another monument to a bowel movement. So unfortunately on both sides of the border, we have zeroes who diddle while two Romes burn and your children, my children, our children are reduced to ashes! I have a new motto for Assr, Sweins and their perversity of diversity, "F...k Hord First, Before They F...k You"! Detroit that has the highest rate of cancer in the entire U.S. and whose population has been decimated so severely, that it has lost a senate seat, is done.

And poor Fendry, dear Fendry, oh Fendry, was sacrificed in vain. They had learned nothing, absolutely nothing from his death or Snowman Hung's death either, for the matter. The same fools were piloting the same ship in the same fashion and running it on the rocks, with all of us on board. I think the appropriate theme song for the 300th Centennial should be the song from the Titanic and we can put Billy Hay Hord Jr. at the helm, steering the ship into a watery grave. Very fitting don't you think, very appropriate under the circumstances. Then we can dress Blow Job Bill and the boys from Archie's bunker as water rats, swarming around the corpses trying to steal any last items of value from the people that they all had helped kill!!

Think Hord and the boys of the Democratic Crypts First, "Only on Autopsy!"

Chapter 27
There Are Only Raisins Under Alex Troutman's Kilt!

Troutman was Hord's Scottish joke, he was supposed to rally the clans, the mercenaries to protect the Scottish honour, aka Hord's honour. Instead he dishonoured them all, entrapping them in his filthy scandals, by giving them free trips and fancy cars, to distract them from the very real life threatening issues that were happening all around them. Mercedes were his specialties, he liked to throw them around to his mercenaries, for two purposes, to buy their loyalties and to make it look like Chrysler's had complicity in this freaking mess too. They paid specific people in the Windsor hospitals to pass on misinformation and spy on the activities that were occurring in them and paid people like Dougie and Bob off with Mercedes, new and used. They even sponsored parties at the Mexican village and other places so they could discreetly pump the staff for information on how much we knew and what information we were sharing. Troutman had Hord flunkies appointed to the hospital boards and administrative staff too. People like Pattie Hurst and Hollywood boy, Paul Cents who thought it was important that the "right" people sit on the boards. That is people who would be willing to cover up anything in order to keep Hord's, Mossad Maxie's and Semi-colon's bottom line intact. Not caring that by protecting any and all negligence here too, that they were sentencing the Windsor hospitals to the same downward spiral that resulted in the deaths of Fendry and Snowman. When people know that every mistake is protected, they don't bother to be careful anymore and carelessness is a very, very dangerous situation in a hospital. Just ask the families of all our dead "die-alysis" patients. They can tell you first hand where their loved ones ended up at the hands of the "Right" people. They ended up Right on the slab at the Wayne County Morgue.

Troutman developed the same amnesia problem that Assr had when it came to safety issues for vehicles, staff and the environment. He was so cerebrally constipated that he needed a king sized dose of milk of magnesia, to clear out the shit he had for brains. Clearly, clearly, clearly the statistics speak for themselves, when the methyl mercury content in the water increased 74% in the last six years with the worst sediment deposits being made in the last few years and when eight babies are born without brains and our incidence of pediatric cancer is thirty percent higher the rest of the country. Very clearly something is terribly wrong. So Alex the Carps, solution to pollution is buying people off with Mercedes' and flying freaking flags over the aquatic carnage that he has helped create. Hats off to Alex another zero, for showing the world how desperate Hord's and their Barclay bum boys are to cover up this freaking sideshow. For his efforts the Carp wins the honorary designation of 00Zero, for going above and beyond the call of duty in this very real Watergate scandal. Alex whose primary focus is protecting Chase Morgan's, First Suisse Boston and Barclay's banks investment in Hord's and making sure that Fidelity Magellan another Hord owner, makes its millions from massacring us. Alex whose Hord vans guard the Vanguard Index Fund shares in the Hord company. Alex gives new credence to the term, "Think Hord First, Only at Auschwitz!"

Chapter 28
What Do the Initials C.F.O. Really Stand For!

I have met many, many chief financial officers in my time as well of perhaps a hundred I would say two are decent human beings. One being Mike Melc who tried very hard to make the creeps from the crypts do the right thing and the other being Bill Donald who also understood that if you did what was best for the patients, the financial deliverables would just come naturally. But it was very hard for them to create balance against the freaks who were so used to covering up everything in order to protect the short-term bottom line and their own personal financial deliverables. Hord was the first place where I learned, the creative "ickonomics" of their chief f...k-over officer. Pom McNutty was a wealthy man, he had made a science of let's make Pom wealthy deals. He knew how to authorize payments on false bills of sale where no goods were ever received so that he could pocket that revenue too. And reimbursement checks from pharmaceutical companies, no problem, he just had them make the check out to himself personally, so he could take care of it, personally. Pom had preferred provider vendors who would make sure he was being taken care of, so they would continue to get Hord's business. He also had friends with sham companies that would bill for a huge amount of contract work and deliver almost nothing, so that they and Pom could split and pocket the difference. I thought Hord was unique in these schill games, until I started working other places and realized that they were all like that and the main objective of the C.F.O. was to screw over the company they were working for, for their own personal benefit, then bail out before they got caught. It was easy to do when they kept the books, one set to show the shareholders, one supposed real set to show the boss and the actual set that showed what the CFO had stripped out of the company, so he could keep track of his hits and not make the pattern too obvious. This third set of books was usually kept on their home computer, like Harry Groandin did, so that no one at the office could accidentally access the files, and spill the beans prematurely.

The CEO David Bumble at PMC was even more creative, he set up external partnerships with companies that were supposed to be in business with PMC, but then channelled all the revenue from the partnership to his own account. No one knew, he managed the business relationship, so the rest of the company had no clue about how much money was coming in and where it was going. Internal audit, forget it they couldn't even figure out that they were paying the same damn bills three times! No wonder the crypts were going broke, they were being milked dry by their own and the yo yo CEO's either were in on it too, or else didn't have a clue.

Their tax scams were even more creative; they split the crypts into several separate business units, but gave only one unit voting interest in the company. To this voting unit they assigned a tiny percentage of the revenue, so that they would end up paying $37 in taxes on the $98,000,000 profit that they had made that year. Not too shabby for creative accounting is it.

The external auditors forget it they had so many of them on the take, that it truly gave new meaning to the Peloitte Gouche douche.

Then there were the living agents, guys who made a living going from company to company, deliberately stripping them out for their competitors. This was a very lucrative business, three different competitor companies might pay a single chief financial officer at the same time, to go in and ravage the business of one, in order to decrease the number of competitors in the market place. Worked well at Chrysler's and GM didn't it and the dummy shareholders couldn't figure out where the profits had gone, even though they were selling more cars and all of these people were supposed to be business people with MBA's. Maybe they should change the acronym to MDA, for more dumb asses, what a joke! You lose $14 billion in a year and you can't figure out where it went! You would think that they would have looked a tiny bit harder, it's not quite the same as losing a sock in the wash, or a buck at the corner store. Wouldn't have taken much for them to find it, if they knew were to look. But now they do, right in Homerica's handbag and it will be fun watching them wrestle Maxie's mossad boys and Archie's semi-colon puppets for the bootie. Ewegene just might find himself hanging from that 31st floor, boardroom window, instead of just hiding under the round table. "Think Hord First" my ass, only when you smell a carp!

Chapter 29
The Impotence of Political Importance!

I have a riddle for you; "How many mayors does it take to screw over the welfare and well being of 1,200,000 people and their loved ones?" The answer sadly enough is just two, two impotent fools who are so bent on building ballparks, casinos and monuments to bowel movements, that they are totally missing what is important. Dead people can't go to casinos, buy cars, go to football games or go out to dinner, at least not if you don't have Bernie's friends. Children with leukemia's, brain tumours, autism and severe neurological impairments, have a real difficult time enjoying parks, when they are too ill to lift their heads. And monuments to bowel movements are no consolation to those whose loved ones are dying of cancer at a rate thirty percent higher than the rest of the country. The only monuments they are left dealing with is selecting the headstone for their loved ones. Perhaps it is time for the respective Hizzoners, to get their heads out of their respective backsides and instead of having Benny and the Jets murder the people that are trying to fix the problem, perhaps they should try a novel approach and help fix it too! Now wouldn't that be a news flash. The same freaks who say there is no problem, despite the fact that there are over 18,900 toxic waste tanks known to be leaking in Michigan alone. Tanks under the freaking fire stations, police stations, hospitals, universities, the works, all the people who have said there is "no problem", while they sat right on top of the freaking toxic dumps themselves. Idiots who say there is no problem, but hey, just boil the water to be safe. Safe from what, safe from the impotence of the politicians and city administrators who continually fail to do their jobs, that's who! Safe from the choices of the "Right People", who would let one of the crypts off their own mother for a dime. Safe from public health officials whose concern first and foremost is the "politically correct" decision to make. Not my God, how many people may get sick and die, but oh bummer will Zero and Hords be mad at us and how much will it cost us. Heaven save us from the political ants who decide to build arenas, when the sewer lines are leaking into the clean water supply and no one does a f..king thing. Or decide to put phosphates in the water to counteract all of the lead that is leaking from the pipes and making people crazy, literally and figuratively speaking. Do they replace the pipes that they have known since 1992, are causing brain damage in the children? Hell no, they are too busy building Homerica Park and Hord Field. Do they support the environmental agencies that are trying to prevent the genocide of their constituents, no way, instead they arrange the hanging of World Health Care workers from the bridge, so that they will mind their own "bidness" and go home, leaving the two respective Hizzoners and their "fambly's" that they have brought to town, to hide the secrets in their respective crypts. Only the joke is on them, because at the rate at which the population is dying off in both cities, pretty soon the only people sitting in Homerica Park will be Mikey and Archie and maybe if they are lucky Maxie will join them too and the three of them can wave their pennants to the crowds who no longer exist because they have been decimated by the Hizzoner's negligence and neglect. What a legacy, two cities with a death rate 30% higher than the rest of both of their countries, something I'm sure they are both really proud of.

Zeroes who diddle while their respective Romes collapse in rubble at their feet. Their creed for the third millennium is one of environmental holocaust, "We take good care of our own and all others we take care of"! Tell that to the 563 children who are dying of cancer in Windsor right now, I'm sure they will appreciate how well the two of you have taken care of them! Assholes!

Chapter 30
Poetic Penile Justice and the Limp Willie Epidemic!

I wonder if all these men who guard the secrets of the environmental holocaust so closely, truly knew just exactly what it was they were protecting, they might change their minds pretty quickly. If they truly understood that the same heavy metals that are sclerosing their coronary arteries, are damaging the arteries in their willies too! That the same neurotoxins from the mercury and aluminum in the water that damage the nervous systems and cause birth defects in the children, also damage the nerve conduction in their penises. Giving new meaning to the term sequins, their crown jewels may sparkle from all the chemical contamination, but lay there as flat as sequins, because the nerves and vasculature are so damaged, that they can no longer get it up. I wonder if Heinz Prickley, Billy Hay Hord, Tony Late, David Baker Schmoois and Ken Wrongway understood the real reason for their failure to go the distance, lay in the glass of water that sits beside their places on the round table of the Homerica boardroom. It is truly a drink up Shriner's toast, because they are toasting their own sexuality with the incendiary chemical stew in the water and they will have to erect (no pun intended), a shrine to their potency, because it will soon be a thing of the past. I wonder if Dickie and Tom Stein understand that their minute man problems are a product of Minimata's disease, you know maybe they wouldn't be quite so mean if their Willies still worked and I wonder if they know that they have Hords and Archie to thank for that! I guess the syndrome can be renamed, "Blue Balls", since the bribery blue boys are making so many men's crown jewels into raisins. But think of it this way, you'll save on medical bills, you certainly won't need a vasectomy, when it won't work anymore and it is so sclerosed, that you can't find it! When your private prize has changed from, "primo al dente", to a little lump of mush, just because you didn't have the balls to tell the freaks to fix the environmental contamination. I kind of think that that is God's plan too, since some men didn't have the balls to stand up to the evil doers and protect their children, God is taking away their balls by the same mechanism. It even says so in the bible, "An eye for an eye, a tooth for a tooth" and I don't think we would be plagiarizing if we extrapolate that to a, "Willie for a Willie." Isn't it kind of funny that so many men named, "Willie", Willie Clinton, Willie Hay Hords, Willie Rice, Willie Hurst would be involved in covering up this mess, it adds a touch of irony to the tales, does it not? Willie's that won't work, because of the "Willie's", kind of funny isn't it! You could really go to town with this if you wanted to. If the performance of your peach is the pits, think Powell! If the cock of the walk has lost his ability to rock, think Vanguard stock! If your sword has been floored, think Hord! If your phallus has lost its' fizz, think Fisher! If Tommy has turned into Tammy think the Tecumseh Company. I think we could give Jason Sweins enough jingles to last a lifetime, but of course they wouldn't help Hord's sales much. They would be like the thunderbird, a lot of noise but no lightening and the way they would love you would be sad, not frightening and the only wood they would be knocking on would be those in their golf bag, because their own willie would have wilted like raisins in the sun from all of the mercury that Hord's dumped in the water. Maybe they should sell Hord cars with jingle bells hanging from the mirror, so that the men who buy them can tie them to their crown jewels so they can find them, after being exposed to the mercury in the ignition switch of their car. If the performance of your peach is the pits, "Always Think Hord First"!

Chapter 31
The Address as Sure as Hell Isn't CBS!

In our search for people to help us with these issues, we met a few more cowardly news people along the way. We had met Mike Willice at a seminar in Ann Arbor where the actions of Jack Kevorkian were being discussed on a panel. We discussed the issues with him too and sent him and Connie Change all the documentation, but that is where big politics came in to play. CBS New York office sits in a democratic state and between the democratic pressures not to embarrass their boy Archie and the black community pressures not to let the world know that it was they themselves who helped do in Snowman Hung. The pressures of the New York mossad community not to embarrass Pisser and Hinai and the pressures of the Chinese community to keep their boy Billy in power . Their boy because they knew that he would sell any secrets to them for a few bucks. That group wilted too, like raisins in the sun. They had lost their raisins when they took on the tobacco industry, their piano players were terrified to address another volatile issue that might cost them a few bucks. Hell it didn't matter to them that the democratic cracker bunker of Detroit was killing us all with contaminated water, and that Detroit had morbidity and mortality rates higher than a third world country. It didn't matter to them that the Canadians in Windsor had a death rate of cancer, heart disease and birth defects thirty percent higher than the rest of the country from the environmental cesspool that they lived in. So the piano players got their way again, even though CBS knew all the details three years ago, even had pictures of all the players and the fire department issues too, they sat on their hands and did nothing. Allowed the carnage to continue like the nants, the news ants that they were. And Dan Rither what a joke, he would rather bias people's opinions to cover up the democratic debacles, so his daughter Robbin, could get a free ride on the democratic meal ticket. Guaranteed, when you help alter public opinion in hallmark cases against Hord's rollover accidents. I would have loved for them to have been at a meeting I attended the other night, a meeting of a small band of people who once again, without a lot of resources were trying to make a difference, to prevent the cancers that had devastated so many of their loved ones lives. I would like CBS to look into the haunted eyes of the young mom, who's three year old daughter is trying to survive cancer and explain it to her why they caved, why they wimped out, why they chose not to pursue the story because of a little political nespotism and pressure. Tell her your reasons for allowing the lives of the 1,200,000 people in this area to continue to be jeopardized and suffer the same fate that her daughter is suffering now. Tell her why it isn't important for the world to know that Detroit has the highest incidence of cancer in the U.S. and that the tiny city of Windsor has 563 children dying from cancer. The address for CBS is filed under the 000 file, joining the other zeroes who knew all this, three years ago sat on their hands and did nothing!

Chapter 32
The Things Tom Peters Never Told You!

I wonder if we can get Tom Peters to write a new book and instead of calling it, "In Search of Wow," call it instead, "Being Pursued By Yikes". A book on how to avoid being killed by corrupt government and corporate goons, once you know their secrets. I think it would be a best seller, given all of the good guys who have experienced the wrath of the democratic crypts and the mossad muppet and semi-colon puppet coalitions. He could start out with the prologue of the ninety some Clintoncides, people who have died in mysterious circumstances when they worked around the Clintons and knew their secrets. That would be a fascinating chapter, highlighting all of those who have died from accidents and suicides in the commerce department. I mean really, would you think that bean counting could be such a hazardous profession! I guess it depends on whose beans you are counting! Then we have the Waco Texas set up, Clinton was a waco who liked to set up both Texas and the FBI and what a coincidence that so many of his former secret service agents that were ready to testify against him accidentally got whacked by friendly fire in the altercation. Was it an act of God, a divine intervention, I don't think so. Try a little help from Billy's drug cartel friends and mossad and I think you'll be closer to the truth! Then he could focus on all the military people that Billy and the boys' offed literally and figuratively speaking. He hated them because they had the guts to expose his government for what it was, a cracker with his mules who would sacrifice anyone's life in order to protect their so-called democratic honour. Those he couldn't have offed literally like General Robertson, Chief of Intelligence for V corp, he had offed figuratively like Admiral Boorda, Chief of Naval operations, through a scandal that wasn't. Clinton and his mules hated the navy's loyalty to George Bush Sr. and made it a point to destroy whoever was loyal to him, however they could. In total there were twelve plane crashes, thirteen murders, ten suicides, four friendly fire deaths and three "accidents", in which a total of ninety key people were killed that knew the secrets of the democratic crypts. Not quite the twelve days of Christmas is it and there would appear to be not a partridge, but buzzards plural in the pear tree! So puleeze Tom, tell me where does one put this chapter in, "In Scarch of Excellence". While you're at it, please tell me where to put the murders of young John and Carolyn Kennedy who were offed because John was going to run for the same New York senate seat that had already been promised to Hillary through mossad and the radical Jewish Hasidic sect, in exchange for the military secrets, sham pardons and other favours they had done for Maxie and his military boys. I can't seem to find the appropriate chapter to place this decimation under.

Next in the food chain of not WOW, but, "HOLY COW", we have the issues of how does a company achieve excellence, when the government of the country is bent on breaking their backs, for bribery bucks. Pretty hard to have excellent financial outcomes, when you are being stripped internally by the crypts, at all levels. Pretty hard for GM, Chryslers, Honda and Toyota to excel, when the democrats are in bed with Hords. Just take a look at who drives which cars and I think you will be absolutely fascinated then take a closer look at what money went into whose campaign coffers before key defence contracts were awarded and to whom they were rewarded. You won't need spearman's correlational coefficient to find the statistical linkages for those events. Unexplained events my ass. As unexplained as Powell and his antitrust boys going after Microsoft instead of AOL, and Texaco instead of Shell, and Coca-Cola instead of Pepsi . It's helpful to have the Jamaican ivory gang at your fingertips, equipped with Maxie's high tech. taping toys to tape people, then set them up by selectively editing the tapes to make them say whatever the keepers of the crypts want, in order to incriminate companies, who won't give in to their schill games. I mean really, who is going to check them for editing, the FBI, who couldn't figure out that Clinton was setting up nine hundred of them for exposure and who walked right into his trap, when he threatened to pardon Leon Pelletier. Or maybe you would like to rely on the CIA who was involved in the Inslaw computer cover up with Billary and Deutch who willingly sold them confidential dossiers on anyone for a buck. How about NASA, to whom the keepers of the crypts sold a bunch of faulty O rings, that resulted in the death of a space shuttle crew. What do the good companies do when faced with such dynamics, when they are being framed into a no win situation by a government that kills off its own secret service agents, who have seen too much!

I'll tell you what the Good Guys do, they align their resources and one by one give the same message back to the keepers of the crypts, not in black and white, but in living colour this time. The Good Guys have made a good start, starting at the top of the tree. The George Bushes squared can play the game better than the cracker democrats had ever dreamed of playing and they do it with a straight face, shaking the hands of those they know are stabbing them in the back, while they let the creeps out on a long enough leash to hang themselves. The Bushes know the mules for who and what they are and have placed them in positions where the whole world could watch their act, knowing that at the first misstep, the Bushes could do an Admiral Boorda on the creeps that had tried to embarrass and destroy the navy. The Bushes know that the mules and the keepers of the crypts are guilty of far more egregious crimes than anything the Good Guys could have ever even thought of and so slowly but surely the evil giants will be killed literally and figuratively speaking, one by one. Then maybe there will be a chance for excellent companies to thrive, not in the fear state of the U.S. Gulag Archipelago, but in the flow state of the truly free enterprise. King Clinton is dead and the Lincoln Memorial is being polished up and those who mocked its values will find themselves like gum on the bottom of a shoe. Dirty and laid out flat for the entire world to see!

THE GOOD GUYS

BRAVEHEARTS, OAKTREE MEN, and WARRIORS

Chapter 33
Team Penske

It is appropriate to start the chapters on the Good Guys with Roger and Team Penske. I never knew who they were before all of this happened, but I clearly understand who they are now and why the freaks in Detroit did everything they could to set him up and destroy them too! I am trying to remember when I first noticed the presence of the yellow trucks. The Hord truck was always there, with it's nose stuck in the bushes at Gino's surf club looking for all the world, like a pig sniffing for truffles. They sat there surfing the airwaves for local calls from anyone who might be trying to help prevent the continued genocide of the children and the poor, weak and elderly on dialysis. They were very concerned, deeply concerned that nothing interfere with their $64,000,000. cash cow, their "die-alysis" business. So they worked very hard to deflect away, anyone that was trying to help. Roger is a quiet man not given to expressing much emotion, but he watches everything, like a hawk or an eagle. When all of this first started happening, he couldn't understand what all the fuss was about either, he just knew that the keepers of the crypts were desperately trying to kill me. Roger who ran his own company, who truly understood the conceptual model of WE, that is the team approach, who supported his Team members no matter how poor and disadvantaged they were. He knew that the keepers of the crypts would murder their own mother for a dollar and that they would hire assassins to kill off anyone who threatened their vertically integrated silos. But what Roger didn't know was that Archie was a Zero too, who hooked the elite on heroin, so that they would support his favourite projects and he also didn't know that Max Pisser was a pimp, who tried to foist off his granddaughter who Tony called the "virgin scary", on anyone who was stupid enough to don one of her pin hole punctured condoms. She was so angry, so enraged at what had been done to her that she wanted to spread more than the word, about aids, literally and figuratively speaking and she did every chance she got. Roger was a Chevy man, a Trucker at heart, who even at 64, looked like a little boy with his toe in the sand, when he put his baseball cap on. Roger who loved to race, to feel the top down wind in his hair thrill of doing 150 MPH on a stretch of open highway or 200 MPH in the closed in cockpit of a formula one race car. Roger who the piano players hated, who tried to reign him in, saying you can't race, no one will insure you if you race, you can't take risks, what will happen to your business if you take risks. Roger, the true Marlboro Man, who rarely showed emotion, but was a man of tremendous depth inside. Roger who saw one of his companies left in ruins at the hands of the son of a so-called friend from Hords. A chippie off the old block, T.D. McWhore engine block, who helped sabotage Roger's son's success, so he could take over the "bidness" for himself. The same so-called friend that tried to frame him at the racetrack by murdering the sons of his real friends, then making it look like he was involved. Roger who they hated because he had more balls than Powell, Patton and Mac Arthur combined. Roger was magic, is magic, because he truly cares. Not charity ball, rainbow connection, do good for social prestige caring, but true caring. From the gut caring, where one puts their heart and soul on the line to stand up for what is right. When even the surgeon general C. Everett Koop caved and the crooked members of the WHO, took their checks and ran; Roger stood tall. When his friends in the tobacco industry took a beating, in order to fill the Democratic war chests with more soft money for their campaigns, he stood by them, not because he agreed with smoking, but because they were his friends. He saw them as real human beings with frailties of their own and he understood that the bad car companies that were killing everyone from cancer caused by the mercury, aluminum and cadmium, were just as guilty as the tobacco companies. Actually even more guilty, because smoking was a choice, whereas the gross environmental contamination that compromised millions of people at a time was more a form of genocide. Murdering all, young, old, weak, strong, killing them without giving them a choice or a chance.

So whenever I was being chased down the freeway, the trucks would appear, to run interference and to block the evildoers from doing me in. The trucks appeared everywhere Detroit, Lansing, Windsor, New Orleans, Boston, Washington and they were always on the right hand side, the side of what is right. Roger knew instinctively without knowing all the details that something was very wrong here. God how they tried to bring him down to their level. Stein, Snalexander and Swines made a religion of it, they even gave it the code name TAZ meant to ruin Team Penske's operation Roger's a hero, to give him credit for what he did in trying to save us all. One of the stupid Stein's even had Phantom of the Opera put on his licence plate. They tried everything, they decimated his son at the motor plant, they set up Laidlow and Raider against him, so they could destroy his trucking company and get his business for themselves. They set him up in a video hyping the city of Detroit, because they knew the story was soon going to break about the fire issues, the police issues, the contaminated water issues, the story that let the world know that the city had no infrastructure, that it was all smoke and mirrors too! They wanted him to look like the zeroes that they were, so they drugged him at the press conference in Atlanta making him look dazed and confused and released pictures to the world press showing Archie's hand covering his face in triumph, showing the world that Roger was in Hord and Archie's pocket. But Roger's son knew, he watched his Dad and saw that he had to get him out of there as quickly as he could in order to keep them from drugging him again. Drugging him this time and killing him the next time so that they could take over his business. When they found they couldn't corrupt him, they turned their efforts to courting his chief operating officer and stripping Roger's businesses through him and the chip off the old Hord block. But Roger and God did a funny there too! Roger sold the motor store to the Germans, causing Chippie to promptly flee and God collapsed the automotive economy so Roger's ex COO's new business venture will go belly up. And now that the Teamster's know who really murdered Jimmy Hoffa, even Jac the Knife's new trucking business venture will go belly up too, because it is the truck drivers who control the deliveries, not the trucks, but the truck drivers who are in the driver's seat and they just might choose to have that second coffee or that second piece of pie, while Hords is falling to pieces, waiting for their just in time deliveries that perhaps will arrive, and then again maybe they won't! And Billy Hay Hord senior who can't figure out why the Lambs can't win no matter what he does, will understand that the ghost of Big Jimmy Hoffa Sr. is sitting between the goal posts on Hord Field, sitting there like Buddha, with a meat grinder in his hands, flicking away like flies any touchdowns the Hords might have aspired to. Now Roger is moving his businesses to Canada, where at least the punks and crooks he has to deal with are small time flunkies. Then all of the evil doers in Detroit and the democratic crypts can go belly up by themselves and Laidlow who thought they had laid down the law and helped cover up all this stinking mess will find themselves driving a little handi-cap bus, because the whole world will know that they and Raider knew that we were being poisoned by the contaminated water all along and they chose to do nothing! Whoever said God didn't have a plan was wrong. That's why Roger wins all the races now, not because Hord is sabotaging the competitions' cars, but because the Good Guys do win in the end, sometimes it takes a little longer but the Good Guys always win! Yea for Roger and Team Penske, our number one HEROES who didn't back down from the creeps from the crypts, no matter how big they were!

Chapter 34
Jack Welched and Sooner or Later You'll Own Generals!

Jack Welch was another player in this pile and for the longest time I couldn't figure out whose side he was on, then I came to realize that for a while, neither could he! Jack was a believer in the principals of quality management and had demonstrated those skills repeatedly throughout his career. Jack had worked for the crypts and knew their tactics well, knew that they would gun down a hundred people in their corporate lobby in order to protect their bottom line then say they had all died of natural causes. Jack knew that that the keepers of the crypts spent all their money on smoke and mirror bullshit then turned their ass to the customer when their cars started to fall apart like tinker toys. Jack had heard enough crap from the boys of the crypts to last a lifetime, so he left and went to G.E. and that's where his dilemna began. His company made mercury vapour light bulbs amongst many other things, so theoretically he should have sided with the keepers of the crypts. But Jack had flunked corporate whore 101 a long time ago too and he was used to running a company his way. Doing it right the first time with no excuses for damaged goods or repeated product failures. So Jack turned his ass to the bad guys, he wasn't going to allow them to murder children through environmental genocide, not if he could help it! Once he did, all kinds of funny unexplained events started happening. Power failures at the city of Detroit building, car windows that wouldn't go up or down, power surges that fried the major appliances of half the city of Windsor, the half that the Hord executives lived in. And after HORD and the freaks from the democratic crypts arranged an accident to hit a young mother who was eight months pregnant and her husband whose spine had been fractured in a HORD-sponsored accident three months before, the democratic enclave of Georgetown was left without power for four days. Funny, unexplained electrical occurrences that sent very clear messages to the bad guys, that if you f...ked over children for a buck, you would be left to sit on your ass in the dark, eating cold beans from a can, so you would have time to think about who you were murdering with the environmental genocide and why!

Jack knew that Assr was a nitwit, who lied through his teeth rather than fixing the flaws in Hord's products that were killing hundreds of people all over the world. He also knew that their primary tactic when shit happened was that if you can't deny it, blame it on somebody else. That's why Jack is roaring now as Hord's takes hit after three billion dollar hit. The gasbags in the glass house have just been blown by their own stupidity. The "No Problem", boys have got big time problems, so big in fact that some countries are going to refuse to sell their cars, because they are so unsafe! Bummer, isn't it, Billy Hay Hord being pimped off to a fanny didn't save their biscuits after all, neither did their linkage to the mossad muppet brigade. As a matter of fact, the mossad muppet piano playing brigade is after Hords now that Assr turned on Hen Hord and the Firestone faction! Too bad, how sad, maybe they should have learned something about doing things right after the Duce's death, but they blew it. And now Jack along with the rest of the Good Guys are having a ball, while the boys of the crypts have theirs in a wringer. Ho, ho, ho, ho, ho, ho, ho the creeps are being fried while we sit back and watch the show!

A fitting outcome for whores who murder children for a few pennies a car and a few bucks on a dividend. Bless Jack's heart for making the "Right" choice and choosing to be one of the "Good Guys". Jack is a Braveheart now too, along with Roger and Team Penske. Bravehearts, Oaktree Men and Warriors, what wonderful guys!

Chapter 35
Electric Al Connections

Violet Al was an enigma. The first time that I met him, he had been sent by the boys of the crypts to embarrass and humiliate me. Easy to do when PMC and Hord owned it's own pharmacy ants like Mustapha too! Just slip some GHB in someone's drink and watch the show.

But a funny thing happened, on Al's way to humiliate me, he fell in love with me and he did because he knew, like every freaking member on the boards of trustees of all three hospitals knew, that I was right. That the reason our patients were dying like flies was primarily because of the environmental holocaust from the contaminated water. And he watched, while the freaks mounted campaign after campaign, to try to destroy me literally and figuratively speaking, both in Canada and the US. Using the resources of the entire freaking democratic party of the United States, as well as the war chests of Hords, Schmer, Homerica, Barclays Bank, First Suisse Boston, Raider, Laidlow, Mossad and many, many other freaks that were complicit in creating the holocaust to begin with.

Al fell in love with me, because even though he was on the other side, he understood, because he had children of his own. He knew that it takes a lot of guts to stand up for what is right, when the other 15,000 people at PMC and 14,000 people at Hord's all laid down like lambs, because they were afraid that the keepers of the crypts would murder them too! Al knew that they had tried to murder me, and then my family, in order to intimidate and bully us into silence. But what he didn't know was that I knew everything all along too and it's amazing what information people give up when you appear to be in a vulnerable position, absolutely amazing how they confirm information from other sources, like Tony Despeisner sources. Sun Tzu strategies work beautifully, even thousands of years after they were written, " In the beginning, be as shy as a young maiden to entice the enemy and lower his defences", then once those defences are lowered and they have exposed themselves for who and what they are, you can whack them on the ass until they are no longer a threat to you or anyone else. It doesn't take money or resources, just intelligence to collapse even the biggest of the evildoers. So Al even though he was on the other side, made the right choice, he chose to fall in love with one of the Good Guys, which saved him from the humiliation and ruin that Hord's and the creeps from the crypts are going through right now. I love Al too in my own way I love the fact that he was willing to take the risk to help, even though it might ruin him. Al belongs with the Oaktree men and Warriors for being a double agent and in the end, choosing to side with the Good Guys even though it cost him many pieces of silver. I guess we should let his stock options go up now and let the Sisco Kid recover and I think Electric Al will be the first to tell you, that he has taken many trips before, but never one quite like this!

Chapter 36
Peter Drucker, Jimmy Hoffa Jr. and the Excellent Truckers

A lot of people don't know that Peter Drucker was a trucker at one point in his life too! Of course some people don't know who Peter Drucker is, so I guess I should explain. Peter Drucker is one of the icons, the gurus of quality management. He is world renowned for his efforts in trying to teach people in all sectors of life how to do it right the first time, so that there is no fall-out, no re-work necessary. No need to recall the Firestone tires if you do it right the first time. No need to recall Explorers, Navigators and Taurus's if you do it right the first time. No need to compromise patient's lives if you do it right the first time. No need to put the lives of 1,200,000 people at risk, if you dispose of toxic waste properly the first time. But unfortunately not a lot of people worry about long-term outcomes, particularly heads of some large companies and investment banks who focus only on today's bottom line. It's interesting though, Trucker's do seem to understand this conceptual model and when you stand back and analyze why, I think it is because they know that one small mistake; one moment's inattention can cost them their lives and the lives of anyone else who happens to be on the road with them at the time. So they become very focused, detail oriented people, masters at specific timing to get a huge truck across a busy highway without getting creamed or to get a load from Toronto to Tallahassee in a constrained time frame, safely for themselves and everyone else. The Truckers don't believe in the Jewish creed, "we take good care of our own and all others we take care of". Truckers understand that you have to take care of everyone, watch out for everyone, even the little people, because one small mistake on anyone's part can cost countless lives. Look at what happened in the fog on 401, one small mistake and carnage. I think that is where Roger developed his understanding too, that every second counts, whether it be in auto racing or deliveries. Certainly in health care every second counts, seconds do make the difference in the survival of a stroke patient from brain injury, or a heart attack victim from a myocardial infarction. I think it would be a good thing to have a few more truckers in politics; they certainly could separate the boys from the men rapidly and give the political ants something to think about beside their own agenda.

I don't think that Jimmy Hoffa Jr. and the Truckers truly understand what power they have in dealing with the evildoers. How simple, how very easy to delay a bunch of just in time crucial deliveries to Hords, the Pissers, the Bunker Brigade, the Democratic Crackers. Just don't deliver the auto parts, the tapes, the CD's, the forms, the paper, the pens, don't deliver the bullets, the computer parts and you put them all out of "bidness". Put all the bad guys out of "bidness", by delaying their flow of resources. Pretty hard for them to deliver their evil messages, when they have nothing to deliver them with, when their flow of resources has dried up like their cash cows. Pretty easy to do, very easy to do if you're a trucker. Easy like Sunday morning!

I also don't think that people understand that the Teamsters do truly care for the little people, but they do and I know that they can be trusted. If you give them a package to deliver to the Justice Department or the Pentagon, it gets there. UPS doesn't have Fedex's selective inhibition, where only pro Democratic mail gets delivered.

I also know that the Teamsters can be a funny bunch too, who can make all kinds of unexplained events happen, like water mains bursting at the Hord Dearborn plant. Hord likes to pee on people, so the Teamsters returned the favour big time! It gives new meaning to the term pee on you doesn't it, when it shuts down the operation of an entire plant! I can see the ghost of Big Jimmy Hoffa really chuckling at that one! And when the evildoers squashed a big truck driver in a loading conveyer belt, making him a ventilator dependent quadriplegic, why shortly there- after a conveyer belt got smaller in a Hord Louisville plant, resulting in the recall of millions of vehicles with cut tires! (that weren't even Firestones). And when the freak of an ex-cop from Horton's gave the young man whose spine was fractured in a Hord-related accident a hard time, why suddenly Timmy's was out $100,000 worth of styrofoam cups. Bummer, I guess Hord should quit going around killing people, especially truck people, Teamster truck people, maybe if they did their luck would change!

Chapter 37
The Domino Effect!

Tom Monaghan is another person who truly believes in excellence. He also had the misfortune of being on one of the boards of the crypts when all this happened. For a long time they tried to snow him too, both literally and figuratively speaking. But his Drivers knew, his Drivers who were out on the streets on a daily basis saw who was following whom and worked hard to understand why. They couldn't figure out what in the world it was that the crypts were so afraid of exposing and the more they saw, the more they told their boss and the more upset he became. He had already begun to plan for his own law school downriver, so frustrated he was with the piano players in Detroit who would screw anybody over for a buck and to keep their evil, corrupt secrets safe. What he didn't anticipate however was that the Canadian Catholics had their own agenda too! The Lemery's were very busy trying to protect the firefighter issues, the honour of the fire chiefs who let their own people roast like chestnuts. The Groandins were busy protecting the "die-alysis" issues of their Detroit relatives; the Dednuts were busy protecting the honour of the bad Drants who had poisoned Heinz baby's baby food. The Cogless Clan too busy protecting the screw up on the new construction at the Windsor Hospital. Too busy allowing nurses to be almost beaten to death, because they were too cheap to hire security guards to protect them. The bad police too busy protecting Dr.Piece and the daughter of one of their own, Shelley Barkhane. Protecting her from what, their stupidity? All this time they had been blackmailed into a fear state, Shelley was innocent, something that they could have found out very easily if they had done a little homework. But no they relied on their friends Sweins, Benny, Wasson and Shexton, who screwed them all over like a cheap suit. Lied to them, lied to the good FBI, everybody. Leaders in education my ass, the only people they were appropriate to teach were the felons in Jackson prison. The rest was all hype, smoke and mirrors and bullshit! And the Polish Catholic honour what a freaking joke, Depuca makes me do just that, he and Clousseau need to grab their balls and do the right thing in order to protect the lives of their grandchildren, instead of sitting on their pompous, hot air filled asses. So poor Tom, the Canadian Catholics screwed him over too! But he is learning, as time and the good Fibbies and the EPA clean up the crypts on both sides of the border, he is learning whom he can trust for the correct information and he needn't feel bad; the locals even lied to the good FBI because they thought they could get away with it. The Democrat's mossad muppets and semi-colon puppets were covering their backsides and so were the keepers of the crypts, so they thought they had it made in the shade, but they were wrong. Because the Good Republicans and Louis Freeh were watching all along and no one can buy St.Louis, not Hords, not Pisser, not the Hinai trust fund, not anyone and that is why the democrats are trying to massacre him too! Tom needs to know how deeply we appreciated his Drivers watching out for us all this time. Protecting us on the side streets where the big trucks couldn't go. Tom knew well the dangers of the crypts, his own Drivers had been murdered for a few dollars and he, himself had been set up repeatedly by the freaks on both sides of the border. Tom also knows that God does have a plan, sometimes it just takes a little longer to deliver, but once the buyers and the sellers reveal themselves for exactly who they are, it's very easy to expedite that plan. Easy like Sunday morning!

Thank-you Tom for choosing to be a Braveheart, choosing to protect the children and those who can't protect themselves . Thank-you for helping us to fight and bring down the evil kingdom of the crypts and the evildoers who would have done in all 1,200,000 of us for a few cents profit per car!

Chapter 38
George Herbert Walker Bush- The Texas Hero

I had done consulting work in Houston and Dallas previously when I had worked for one of the crypts, so I knew that the Texan's truly did care about their people. Even at the height of the oil economy recession in Dallas, they still implemented the heat sterilization process for the dialysis patient's artificial kidneys, so their patients and staff wouldn't be exposed to the cancer causing toxic agent formaldehyde. They didn't want formaldehyde getting into their water supply either, they knew that it would sicken and kill not only their people but their cattle too, another key part of their economy.

The Bushes couldn't understand what all the fuss was about either; they just knew that the cracker democrats hated us with a passion, so they tried to investigate too, into what was happening. They didn't know the history and complexity of what it was that the keepers of the crypts were so desperately trying to hide, didn't know who murdered Hoffa and that the contaminated water issues that every freak in Michigan knew about at least ten years ago was what had done poor Fendry in. The blacks were desperate to keep hidden the fact that it was they themselves who were complicit in offing Snowman Hung, because they were responsible for the water problems. But since the origin of the trail of the contaminated street water lead to Archie's bunker, they pretended to be blind mice too! Chasing the ghosts of the Mississippi instead of dealing with the very real issue, that in the shadow of three of America's largest health care systems, their own people were dying like flies. Clod Hung knew, we had gone to him repeatedly for help. The Southern Christian Leadership Not Council knew. I have a copy of their report printed in the 1980's, that details the morbidity and mortality issues extensively. They all knew but chose not to do a f..king thing and as a result endangered not only their lives, but that of all the Canadians on the other side of the border. The Bushes didn't know all that, they just knew that the Democrats, the keepers of the crypts and a few republicans who could be bought like three dollar hookers too, were desperately trying to cover something very, very big up. Like the real reason why the death rate for cancer, birth defects and heart disease here was thirty percent higher than the rest of the country. The keepers of the crypts knew why, that's why they paid the cancer specialists, the cardiologists and pediatricians to falsify the statistics on the true causes of death in this region. Easier to throw a few cheap leases and a few free trips at the Drants, than to fix the fact that they were using the Detroit River for a toilet daily. Please understand, I do know many wonderful physicians, but unfortunately I also know some Drants. Tony Despeisner was clear and specific about whom they could buy like a cheap suit, so unfortunately I knew them all!

When the Bushes tried to help us, we came up against some very big obstacles. The biggest being code 88, where federal agents cannot do or say anything against the current president of the U.S., even though he may be the biggest boil on the ass of the world, literally and figuratively speaking! And boy was he ever; they called Blow Job Bill, Gore, Lieberman and Powell the gruesome foursome. There seemed to be no limit to what they could come up with and no limit to how tacky they could be. They planned all kinds of OPS (operations), to route their opposition and create distractions. Like planning the murders of young John and Caroline Kennedy, because he was too charismatic and spoke the truth about their political shortcomings and because he was going to run for the same New York senate seat that had been promised to Hillary by their mossad friends. And arranging the murder of Trudeau's son to give the Canadians something to think about other than Blow Job's embarrassments, like the fact that he gave Aids to the Canadian haemophiliac children by selling us contaminated blood products that the U.S. wouldn't touch. They enlisted the black mayors' and the Jesse Jackass coalition's help to identify companies they could sue for "civil rights" violations like Texaco and Coca Cola, all the while ignoring the genocide of the dialysis patients that they were profiting from. They picked their targets carefully, crucifying the tobacco industry for causing cancer, while knowingly and wilfully ignoring the fact that the Hords were doing the same damn thing, only with mercury and causing a much higher death rate. They lined up David Noise, who likes to molest little boys up against Bill Gates, to provide a further bonanza for their campaign war chests. That was a sweet deal; Boise got 25%, which he promptly invested in Napster, a company that competed with Bill Gates! And of course Michael Powell who was in bed with Booby Pittman of AOL Time Warner made sure that it was Microsoft that got decimated, so his stock options would go up. Talk about securities fraud and insider information, the whole stinking Democratic government was one big fraud, a sham, a schill. But what do you expect from a group that used the White House as a sheets, no sheets motel and a democratic "justass" department that bought Janet Beano and Mike McDirty out for a cheap lease on a Hord! They would sell access to top-secret information to anyone who waved a few bucks under their nose, or who showed them their bare bum. They even rented out the Lincoln bedroom for Christ sake and did little teas for anyone who would pay the appropriate price. Coffee, tea, muffins and munitions, whatever it took to keep the silver coming in, they would sell.

Even when the Bushes' won, the freaks wouldn't stop. They did little let's make a deal with McCain and his mossad muppet partner, so that the Bushes would have to fight not only one shadow cabinet, but two. The ticked cracker democrats and the disgruntled republican wanna bes. Pisser and the jewish coalition were just dying to get mossad representation in the White House, so they could fight the Holy war for the jews and it wouldn't cost them a thing. So they backed Bushes' opponents to the max, they lined up all their resources against him, the resources of the Pissers, the Hords, the Jewish coalition, the Jesse Jackass coalition, the cracker democrats, the southern McDirty's and the disgruntled republicans, and they lost! They lost because God does have a plan and it doesn't include those who murder innocent people, whether it be soldiers, federal agents, tiny babies or disadvantaged people on dialysis. God has a plan and it doesn't include freaks like cracker democrats, Pisser, Hords and the Bunkers, that sell out their country for a cheap lease on a Hord and thirty pieces of silver!

Chapter 39
The Fabulous Fibbies!

Talk about a stinky job to have! You get to live out of the trunk of your car, survive on fast food, do your laundry in laundry mats, drive subcompact cars and be paid minimum wage, all for the privilege of being shot at, not only by the other side, but your own side too! No wonder so many of them, like Mike McDirty were on the take in the big cities. It really was a stinky job, but a fascinating one too and how my respect has grown for the Good Fibbies over the last three years! What a job to have to worry about not only the enemy but your own local bureau chiefs setting you up for thirty pieces of silver. What a job, what a lifestyle to deal with the plots, the sub-plots, the sabotage and subterfuge on a daily basis. It's like being trapped in an eternal Stephen King novel, that just when you think you've seen it all, another rat comes scrambling out of the woodwork to bite you on the bum. Not little rats either, BIG RATS, GREAT BIG RATS, GREAT BIG RATS on your own side too, that make the Mafia people look like cub scouts in comparison. What a challenge, to try to fight the bad guys, while the supposed good guys are keeping track of your every movement. When your own Government conspires with the banks and big business, limiting your cash withdrawals to a hundred dollars at a time, so they know who you are and where you are. When your own Government arranges for key agents to be infected with boils, so that they will have to surface periodically and seek medical treatment no matter where they are. When your own Government and sometimes your own local bureau chiefs sets up sham pardons, operations and ruses so they can pick you off one by one, at least nine hundred of you. What a life, to have to worry if someone hugs you that they are not going to slap a transdermal hallucinogen on your back, or poison the coffee you pick up at the local donut shop. What a life to have to worry that the pork chops that the supposed good guys feed you is laced with pentobarbital, so they can capture you and torture you with electric cattle prods, cigarette lighters, whatever, to reveal how much you know and exactly what you do know about the keepers of the crypts. But the good Fibbies do it, they do it every day, to keep you and I safe from becoming victims of the same horrors and atrocities that they are subjected to on a daily basis. Just think of living a life where every time you go to bed at night, you wonder if you will be alive in the morning and every time you start your car will it blow up and every time you lose a friend, will you be the next one to die in the line of duty or become a MIA (missing in action), at the hands of your enemy or one of your own. They must have a fascinating job description under other duties as assigned. Must be willing to hang out in women's bathrooms to hear all the local scoop, must be willing to dress up as ladies, he-males, she-males whatever, in order to protect their identity. Must be willing to wear flak jackets under your clothes, so that when the local police or your local bureau chief shoots you, you have a better chance of surviving. Must be willing to hobnob with bureaucrats then dress like a street bum the same day, in order to get the information you need to keep the people safe. Must be willing to never look up, so that the supposedly friendly satellites can't track you. Must be willing to scan your clothes every day to detect tracking sensors they might have placed on you. Must be willing to have your personal life scrutinized, bisected and dissected both by the enemy and your own side for signs of weakness that they could use against you or your peers. Yet they do it, the GOOD FIBBIES, they do the job anyway and they keep their humanness, their sense of humour and their ability to laugh and cheer when the Good Guys win. Which creates balance for when they lose one of their own. The GOOD FIBBIES are very, very special people, they make the marines look pale in comparison to their valour and Louis Freeh even though he has tunnel vision sometimes (literally and figuratively speaking) makes Powell look like a piss ant! And this message is to Mike McDirty, please tell Tony Despeisner Hi, the next time you see him. I'm sure he'll think it's funny that you made him one of the ghosts of the Mississippi too. And Mike it is not very attractive for a high-ranking FBI official to lie on Lincoln's birthday, there is a new Government in the White House, one that doesn't sell Lincoln's bedroom, or hadn't you noticed! Not very attractive to delay the exposure of the FBI Agent, the Mortician, until Bush was in the White House, to save the democrats the embarrassment that he was doing the same thing that Billy and his cabinet were doing, selling all your country's secrets to anyone who would pay the price. Mike do you know why they call me Peaches? It's because I'm an expert at shaking the trees, no matter how big they are and making the rotten fruit fall out! Two Bushes are worth all the democratic crackers in the world Mike and you showed the world you're not in George Herbert Walker's Bush, Mike or in the FBI's bush either! Shame on you, you could have been a Braveheart and now you're just another plant!

Chapter 40
Harvey's Rabbits Aren't The Only Thing to Disappear!

Another miracle on Maple Road happened today. Harvey ended up not being the only piano player whose rabbits disappeared! We inherited another missing in action in the Tales of the Crypts. This time it's Tony Despeisner, whose son works for Al Gore, who thought he ran the FBI, or maybe he does run at least one flunkie there. Mike McDirty has no knowledge of, can't find the father of one of Gore's staff, the historian of the Tales of the Crypts par excellence. Mikey can't find him, or perhaps Mikey doesn't want to find him, because Mikey is either a) complicit in their schemes or b) doesn't have the guts to address the issues. My vote goes for a) since Mikey had me spell things out for him a long time ago, when all this began, when he was investigating the issues for the cracker democratic federal government. "Tell her to spell it for me" was his exact wording, but when they found out what a mess it was that we had uncovered and we understood who was involved in murdering whom, Mikey and the cracker democrats got cold feet. So now Mikey has developed amnesia, or perhaps he needs milk of magnesia, to help him clear the shit out of his brains and improve his memory. Not very attractive Mikey, not very attractive at all that you can't find Vor, the God of the OR is what Tony called himself because he had covered up every surgical mishap that had occurred in Detroit's operating rooms for the last thirty years! I have some suggestions for you Mikey, look in the Royal Oak phone directory under 248-644-3007, look at 1842 East Maple Road, look at the Caribou Club or in the yellow pages under medical malpractice defence attorneys. Or better yet, ask your friend Anthony Despeisner Jr. who lives right there in Georgetown where his dad is, or maybe his good friend David Boneyard. Perhaps then your success rate in finding the disappearing rabbit might improve. Not very attractive Mikey, I found you easily, out of all the FBI agents in the world, I found you just fine, just like I did Tony, now I would suggest that you do your job and find Tony or the George Herbert Walker Bushes and John Ashcroft are going to want to know why. Mikey, Mikey, today is

Lincoln's birthday, we gave you a second chance and you blew it one more time. Not very attractive Mikey, not attractive at all!

May 2001 Postscript : Isn't it amazing that all the documentation on the McVeigh case that was ordered to be withheld by Clinton and Janet Beano, is suddenly appearing! Amazing isn't it, when you finally have someone in the White House Justice Department who does believe in true justice, amazing what can happen. I bet John Ashcroft can find Tony Despeisner and polygraph him too! What a confession that will be, it will make Tim McVeigh look like a choirboy in comparison. The true confessions of the democratic crypts, it will be a blockbuster hit. Complete with sex, lies, multiple murders and videotape. It will keep the Republicans in office and the Democrats in the can, forever!

Chapter 41
Yea, Yea, The W.H.O. and The E.P.A.

Probably the next most thankless jobs are those of the World Health Care Workers and the Environmental Protection Agency. People who think it is a cushy job, are wrong! Not very attractive, when a bunch of corporate whores try to hang you from the bridge, because you're digging into the sediments on the Detroit River to understand just how bad the mercury, aluminium, PCB and other toxic waste contamination has become. Not very nice when those self same whores poison your food, threaten your family and force you to work under cover, just like the FBI, so you can stay alive to do your job. Not very nice, when your own side, the very people you are trying to protect, hate you because they think you are destroying their jobs. Destroying their jobs, my bare butt! The same money the evil corporations could be spending on installing scrubbers in their smoke stacks and treatments plants for their toxic wastes, they are blowing on hiring assassins to kill the good guys. Why spend a couple of million to install a scrubber, when you can hire the Boycano Brigade, the Jamaican Ivory Gang, the Moxley Crew or part of the Rabid Radical Rabbi sect to off the Good Guys and send them a message at the same time. Murder the WHO, the FBI, the EPA, The Hague, the UN, hell it doesn't matter to them as long as the profit margin is good this month. So the black blob grows in the black lagoon and Hord flies f...king flags over their leaking oil booms, to keep the birds from eating the poisoned fish, so all the birds don't die too. So they can continue to let the contaminated petroleum products leak into the river and kill the fish and kill the people too! No wonder the W.H.O, the E.PA. and Amnesty International had to set up permanent offices in Detroit, it was so they could counter act Hord's solution to their pollution. Flying f...king flags to scare the birds away, while they continue to poison us all, with their PCB's and methyl mercury, two of the most cancer causing toxic agents known to man. Let Hord's tell the parent's of the children with rare leukemia's that live here about Hord's solution to pollution, their f...king flags and then come back and see me. I have an appropriate joke for you that the WHO and the EPA will enjoy. "How do you tell the difference between a rectum and an a..hole? The a..hole has a Hord emblem on it!" What fools, what idiots, what morons, now you understand why we think Hord first, particularly at autopsy!

And whenever there is an autopsy on people like Jamie Barker, who knew about the World Health Care people who were working in this area, we know exactly where to look, across the river at Archie's bunker and corrupt "Big Bidness", who always put their few cents on a dollar profits before the welfare and well being of anyone, even babies and children! Fidelity Magellan, Vanguard Index, College Fund, Barclay's bank, Hords, Assholes!!! God bless the W.H.O. and the E.P.A. for being Bravehearts and having the guts to stand up to them!

June 2001 Postscript: Amazing, isn't it, the city of Windsor that continues to cover up mossad muppet murders for "Big Bidness", has been hit by another plague. This time, it's strachybotrys mould that has infected twelve offices in Windsor, who were in collusion in covering up the fact that our people are dying like flies from environmental genocide. Even the "Pubic Health Office" (a real joke), is infested. I wonder if they get the hint, twelve offices for twelve tapes, that tell in the mossad democratic crypts' own words, who is complicit in all the murders and why! Do they understand how easy it is to create a plague of epidemic proportions over night! Ask Barclay's Bank Britain how devastating an epidemic of mad cow disease can be to a country's economy. Then think of how easy, how every easy it is to plague a corrupt city forever, with out breaks of mould, mad cow disease, legionella, e-coli, the works. It's time the city of Windsor learned to straighten up and tell the truth. They chose to become purgatory in accompaniment to Detroit's Hell Michigan and the only way they can get out of purgatory is to come clean. Otherwise, the curse of the crypts, will continue. Canada is not immune, the U.S. economy has lost 666 billion dollars in the last two years, the same can very easily happen here. An eye for an eye, a tooth for a tooth and a billion dollars in losses for every Good Guy they have murdered, perhaps then they will learn that there is a price to be paid, when one colludes with Hell.

Chapter 42
The Wonderful Waltons

Whatever damage was being created by the cracker democrats, seemed to fall to Sam Walton to undo. What a job that was, everything that was good and pure and wholesome about Arkansas went right down the drain, after Blow Job Bill and his Missus had passed through. Poor Sam had a full time job running across the country in his beat up pick up truck, helping people to understand that Arkansas wasn't all that way. That there were people there who did believe that people do make all the difference and who tried to treat others with fairness, courtesy and respect. Sam, like Roger understood the conceptual model of "We", that it is the spirit of the team that makes a company soar. Not bullying, harassing or intimidating people into silence. Not murdering them or their loved ones or locking former business partners up in jail so they can't tell. But nurturing, guiding, loving and supporting people to make them shine and blossom and grow. How different, how very different the styles of management and how very, very different the outcomes. The Good Guys thrive and the bad guys lose billions of dollars in less than a year, because they made the wrong choice. The bad guys chose to bully, harass, murder and intimidate the wrong people this time and now they are learning to be held accountable for their actions. A lesson that they should have learned a long time ago. Sam Walton is a hero too, as are his people, they watch and they learn and they listen quietly. They know the good from the bad and the ugly. They knew that it was some local creeps from the crypts who tried to set them up and screw them over by creating a phoney oil spill at a local store, to make it look like they were neglecting the environment too! But Sam Walton doesn't have carcasses of dead birds and fish laying around the outside of his stores like beacons, leading the E.P.A to the true trail of who is murdering whom and why. Sam Walton doesn't fly freaking pennants over his oil booms to keep the wildlife away, to keep them from being poisoned. But other people do, actually all the world knows who would throw six million dollars at the wall in three months to make a football team win, then fly f..king flags over the carcasses of the wildlife, that were killed by the same contaminated water, water that they contaminated, that is killing us all. When deaths occur in Detroit and Windsor now, we think Hord, the cracker Democrats and Archie's bunker first, especially at autopsy. Assholes!

So Sam, another Braveheart will continue to thrive and grow, while Archie Bunker, the Pissers and the Hords crumble around him, because he does understand that GOOD people do make all of the difference!

Particularly once they know, that what is at stake is the lives of their children and their loved ones!

Chapter 43
The Crucifixion of Chryslers

The Rabbid Rabbi Jewish community were excellent at making things disappear, not only Lady Di and Despeisner, but also billions of dollars in revenue. When it had become clear that Schmeaton had mismanaged Chryslers into the ground, they had to find a buyer big enough to absorb the mess, one who might not look too closely at all three sets of their books in their rush to obtain a foothold in the North American auto market. So Schmeaton and their chief corporate council Lionel Ritchie hatched a plan to save themselves some embarrassment and secure some very golden parachutes at the same time. Revenues were down a little, so what difference did it make to the soon to be retired American Corporate team, if they raped the company a little more. Stripped out the revenue by paying phoney bills to shell companies, their shell companies in order to guarantee their parachutes were really golden. All the mossad muppets and semi-colon puppets helped, Pisser and Homerica, along with Homerica board member, David Barker Schmoois of Schmoois and Stunday, whose attorneys handled Chrysler's legal affairs, were all in on it! They were wizards at setting up sham companies, Archie and Stein of the local democratic line, helped too! Their logic was that they were American, so they had a right to screw the Germans, their public line, Ewegene's line, Stunday's line, Swein's line to cover up their real motive, their underlying greed. Their motto, "We take good care of our own and all others we take care of". And so they, the jewish and the black elite, with Hords help too, stripped out the profits of Chryslers slowly, meticulously over the process of two years. Did a Sunbeam on them, setting up sham companies to whom they bled off all the revenue, through sister-in-laws, brother-in-laws, friends, numbered companies that only existed on paper. And miraculously, they made all the revenue disappear, just like Harvey's disappearing rabbits. They knew the whole time that they were destroying the company, which would result in shutting plants, putting thousands out of work, probably putting some even out on the street, but they didn't care. "We take good care of our own and all others we take care of". More "bidness" for Hord, more money for the mossad community causes and the evil black monuments to bowel movements . They didn't care that they put yellow, black, orange, white or pink people out of work as long as their pockets were full. This was another accomplishment that Despeisner's team was so proud of, proud that in their minds they raped the Germans. Choosing not to see, selecting not to see, hoping no one else would see all of the good people they were destroying in the process. No wonder Despeisner disappeared, because once they caught him, they would hang him for all the atrocities they were involved in and it wouldn't take much to make him tell, just a cheap lease on a Hord and thirty pieces of silver.

But God has a plan for Chrysler's too, a recovery plan, because Chrysler's doesn't use mercury in it's ignition switches, it doesn't poison their people and the environment with one of the most toxic agents known to man. So Chrysler's is one of the Good Guys too and wouldn't you rather buy or lease a car from a company that wasn't killing you and your family. Most people would! They just need to let the world know too!!!

Chapter 44
GM Gets Jimmied

I feel so sorry for GM they tried so hard to help too at first before Hord's corporate piano players made them nervous about the Continental Aluminum plant that supplied them both with aluminum ingots. Everyone knew that the plant was an ecological disaster that discharged its effluent right into the local water supply and asphyxiated it's residential neighbours with the oxidizing fumes from it's plant. Everyone knew, but GM wasn't responsible for treating the water before it came back to the "die-alysis" patients, but Archie, Hord, PMC and Hinai were. So in typical corporate whore fashion, the bad guys sent their piano playing clowns over to the Good Guys, to create a panic state, a fear state, just as they had done to Chryslers. Yes I know that GM does need to get rid of the mercury in their ignition switches, but you have to have a little compassion for them. Just think if you had to work at the top of the Renaissance Center in the city of Detroit. A city with no police, no fire, no electrical services to speak of, no sewage system, no clean water, no infrastructure to speak of. And you had to sit there drinking the poopy water that you know had almost solid waste in it and you had to think about the fact, that if there was a fire, they would all be done like dinners. Jack who circled the Wagons, their new finance boy from Hords; who is screwing them over too, just like he had quietly helped Schmeaton screw Chrysler's. Their new finance boy knows that double dipping can be very profitable he's done it before. So I feel kind of sorry for poor Jack who had circled the wagons, he was had three years ago and still is being had now, by the same players, screwed without a kiss, by the same players. I was going to send him a "Detroit" survival kit as a funny, with a little plastic police whistle in it for blowing the cops, just like they blew him and a little coffee filter for straining his water and a little red toy fire truck that he could play with when his building caught fire! But I didn't because in reality it is not very funny at all. Every day more people disappear in Detroit and nothing happens, every day hundreds die from super infections from the contaminated water and nothing happens, every day more lives are lost in the city of Detroit from fires because the firemen have nothing to work with and nothing happens. Zero truly does diddle while Rome burns. So I would suggest that Jack concentrate not on a golden parachute at this point, but perhaps a silk one with a rip cord, or maybe a hang glider to get him safely down from his lofty perch, before he becomes the next victim of Archie's charade. Jack thank you for Keith, his presence was very reassuring to me, he went through a lot too, when they tried to kill him by running him off the road. I just feel badly that you listened to the wrong people and that you bailed out too soon! Jack there is a funny in here for you too, your new finance boy screwed Hord over royally before he left, just like he's doing to you now. It's very easy to authorize payment of very large non existent bills to be paid to sham companies that you have partial ownership in, when you're the CFO, very easy indeed. Very easy to make the books look good when the only thing that is left are the bones, just ask the boys at Sunbeam. Easy like Sunday morning. The funny part is why in the world you, Harry and the rest of your wizards were never able to figure it out. Do the math and check your finance boys' personal books, I think you will be amazed! Happy Valentine's Day Buttercup-Luv Peaches.

Chapter 45
The Honda, Toyota, Nissan & KIA Honeys

Probably the nicest thing about the Japanese is that you don't have to explain to them about who Deming, Drucker and Ohmae are. They know instinctively what brought them out of the recession after World War 11, was not only their penchant for hard work, but the clear understanding of what these three men taught. That you always do it right the first time and that if you pay attention to the little details, you make a product that never disintegrates and people come back to purchase that product time and time again. No "Fix or Repair Daily", or "Found on Road Dead" acronyms for their products. They like Roger focus on precision engineering, but good precision engineering, not sabotaging the competition's cars, but winning by consistently making quality products and making every second count. Honing the skills of their teams and imbuing them with a sense of pride in their accomplishments. The Japanese know all about corporate espionage and watch in fascination while some American companies focus not on making a better product, but on killing off their opponents, literally and figuratively speaking. And while the evil doers like Hord are busy killing everyone off, Honda, Toyota, Nissan and KIA continue to increase their market share, because their cars don't roll over and their steering columns don't fracture and their cars don't stall in traffic on busy freeways. So the Japanese were sympathetic to our cause and protected us on the side streets too, where the big trucks couldn't go. They knew all about methyl mercury poisoning first hand from their own ecological disaster that had cursed generations with Minamata's disease. The mercury contamination of the water supply in one of their ocean ports had caused countless cases of birth defects, mental retardation, cancers and early dementia. That's why Honda, Toyota, Nissan and KIA like Chryslers (at least the Canadian Chryslers) don't use mercury in their ignition switches. Honda doesn't because they know the terrible cost in human life that is paid for such inexcusable negligence. So the Japanese car makers are Heroes too, because they knew and they tried to make the keepers of the crypts pay attention as well. But our adversaries were huge, with tremendous resources, pretty hard to fight a battle against people who had the president, vice president and attorney general in their pockets. But things have changed, there is a new president on the block and a new Attorney General who doesn't agree with the philosophy that vertically integrated industrial juggernauts can kill off anyone they want, in order to keep their thirty pieces of silver in their own cookie bags! A new government, who is airing out the Lincoln bedroom and making sure that those who stink of corporate corruption, never get to sleep on it's sheets again!

Chapter 46
The Marvellous Mafia People

I don't think I had ever met a mafia person until all this happened and once it did, boy I met lots of them. But I learned there too, that there was good and bad in their clans as well. It was funny in some ways that the government was always going after them for their involvement in money laundering, gambling, racketeering and prostitution. Now really, give me a break, what do you think the city governments of Detroit and Windsor are involved in, it certainly isn't making rosaries! So on one hand, we see all this tremendous investment in resources to capture a mob boss who ordered the killing of one person, while we have Benny and the Jets guarding their body repository of eight hundred bodies at the Packard plant, and hiding the fact that Detroit currently has over eight thousand missing persons, and no one does a thing. Boggles the mind doesn't it when you think about it, no logic at all as to who they go after and why. I am grateful to the Italian mafia people, because when Archie's bunker was so desperately trying to kill me, with the carbon monoxide and punctured tires, they watched out for me too. Every day, checking to see if I was okay, because they knew who the freaks were, who lived above me and knew the relationship of the person who managed the building that I was in, to Hords. Then once I met Tony Despeisner, I learned that there was a Jewish mafia too and they were not nice people. They loved to commit murders and get away with them, the more atrocious the better and then blame it on someone else. Like the murders of Louis Freeh's friends in Italy, the Giovanni's . That was the work of the Jewish mafia, but they did it in Italy, in the cradle of the Italian mafia, to deflect the blame and make Louis Freeh chase the Italians and not like them. Like the murder of Lady Di, they did it in France; to make sure, ensure that no one would pinpoint them. The murder of the Kennedy's, they arranged that to occur in the least Jewish local in the entire United States and then they laughed and gloated, because they got away with it. Killed Kennedy in order to give mossad sympathetic Hillary the senate seat, in the jewish borough of New York. They got a huge kick, a huge thrill out of planning "ops", accidents that weren't accidents to get rid of the competition or to get rid of people who shunned them, or worse yet, threatened to expose them, like the World Health Care workers, that they had Archie hang from the bridge. The jewish mafia revelled in the fact of how many people in high places, they had in their pockets. Certainly Billy Boy Clinton, Gore, Lieberman and Powell were their boys . They had enough goods on all of them to keep them in line and have them espouse whatever party line big "bidness" wanted them to. For the local mobsters, the keepers of the crypts arranged accidents that were not accidents for them too. Like killing the kill switch, on a boat so the engine won't shut off, just like they had done to mine the year before. Or arranging for RCMP flunkie Barbar Boinko to spot one of them at a local casino, to put them back in the "hoose-cow" or arranging to cut an oil pipe to damage their crops and "create" an environmental disaster to keep them in line. The jewish mafia also made sure that the Drants they had in their pocket, didn't find John Gotti's cancer of the throat and mouth, until it was too late to save him! No job too big, no job too small, the keepers of the crypts and their mirror image flunkies do it all!

Call them A framers or D framers, call them mossad muppets or semi-colon puppets, call them what ever you like, but soon the keepers of the crypts with be F framed, for federal penitentiary framed and there will be new pictures on the wall. America's most wanted pictures of all of them in black and white!

Chapter 47
Raubman and the Retailers!

Al Raubman was probably one of the funniest guys in this whole dynamic. Al was a bandit, a Robin hood, who didn't have a lot of formal education, but who was able to parlay his innate business sense into a retailing empire. Al hated, despised the holier than thou, old money aristocracy, the A list, who had F values when it came to treating other people fairly and decently. So Al made a game of it, jacking up the prices for the so-called princes and princesses, who liked to denigrate him behind his back. So he took them to the cleaners to their fronts. Charging them outrageous prices for their designer wear, that he had bought at discount pricing and charging them exorbitant leasing costs in his malls for retailers like Heman Markup. He even anted up the pricing to the super elite, who purchased old world art through Mothbeys and Cookies, and he roared, he laughed and he had fun with it. What the so-called A list valued anyway was how much they had paid for their works of art. Very few valued the beauty of the pieces themselves. They liked to whisper quietly to their friends, that this is a "Dega" and this is what it cost. So for a long time Al got away with it and passed along the wealth he accumulated by giving discount pricing to the poor in areas that didn't have a lot of money. The big joke was that you could buy the same evening dress for $99, in one of his discount stores, that the super rich paid $9000 for in one of the more upscale stores. Al would have continued to get away with it too, if it wasn't for Despeisner and his democratic connections. Al had been trying to help us too because he knew all about the cesspool the freaks in Detroit were trying to cover up. So the Homerica whores with Pisser's and Despeisner's help, bought off a couple of whores, one from New York and one from Dearborn, to set him up in stings both large and small. The democrats figured he would be an easy target to milk for soft money for their war chests, while at the same time, currying favour with the miffed elite that Al had robbed. Despeisner had called it bagging two sets of birds with one stone. The birds being the pigeons that they suckered into funding the democratic crypts and the lifestyle to which the piano playing freaks had become accustomed. So they took Al to the cleaners, but Al did a funny there too, he gave the feds a little insider information of his own, a little poetic justice so to speak. Now every time certain companies burp, the Feds are all over them like white on rice. They turned him in, so he returned the favour, see the Good Guys do win in the end. Al thank you, I know who protects my oldest daughter and who has protected her for the last two years, ever since that freak of an ex-cop who robbed the Horton's, tried to set her up and do her in. The same freaks who arranged to have the pipes cut over an electrical transformer in her store twice and tried to set the place on fire twice through accidents that are not accidents. Messages from Hord, Charles Wasson and Benny and the Jets, who were miffed that the Feds were all over them like flies on a poop pile. Al is one of the Heroes, the Bravehearts, the Oaktree men, the Warriors. He didn't lay down for a few freaks who are finally realizing that their absolute power exists only in their minds. Assholes!

Chapter 48
The Magnificent Microsofties and the Snoracle Debacle

The Microsofties are wonderful people who also tried so hard too, to protect all of the people in this area, but they had quite a trilogy to deal with. Snoracle was in bed with Hord in developing Hord's information systems, both at the hospital and their automotive holdings. Snoracle had Billy Blow Job in their pocket, literally and figuratively speaking. They had intercepted a lot of the golf course conversations of the gruesome foursome, conversations that it would not be very attractive to make public, like who they were going to off next and why. So Billy Boy ever receptive to bribe and coercion agreed in exchange for some funding of his own, to set Janet Beano on the tail of the Microsofties, who were bad for Snoracle's "bidness". Another case of triple dipping, they got to destroy Microsoft's reputation, deliver the business to Snoracle and AOL who were light years behind Microsoft in their technology; probably because they spent all their time on corporate espionage instead of product development and thirdly to fill the democratic war chests one more time. But with Bill Gates money this time, so they could cripple Microsoft for Snoracle and AOL and make Billy Boy a shoe-in at the same time. David Noise was a willing participant, because he got to keep 25% of the spoils too, on a case that he was sure to win because that is what Clinton had told Beano and Michael Powell of the "justass" department to decide, so they could all keep their jobs through the next election.

Securities fraud was taken to a new level with the cracker democratic government, who set up and framed any of the good guys who didn't support members of the crypts. Michael Powell who worked for the justice department made sure that his friend Booby Pittman and AOL got all the "bidness", that they were stealing from Microsoft, to make sure Mikey didn't have to work too hard in future. So don't be surprised if you can see a few more AOL and Snoracle favours tossed Billy Boy and Beanos' way, like free leases on Hord cars, a little help with the mortgage payments and maybe a high paying board membership or two, or maybe they will sponsor them as speakers, at $100,000 gigs, in places like Hong Kong so they can continue supporting their boy through discreet, oblique financial ventures, so he can leverage his democratic friends and his wife who is still in power for a few more favours in the future. But the democratic cookie bag is being rattled now, the Republican's don't like former president's who set up their FBI agents to be murdered and sell their country's secrets to Israeli spies. They don't like democrats who sell their country and the welfare of their people like the three dollar hookers that they are, so now the House, the Senate and the New Justice Department are demanding answers and life will never ever be the same for the cracker democrats and the keepers of the crypts again. Bummer, I guess next time they will think twice about murdering poor people through their negligence and exposing people on both sides of the border to premature death from the environmental holocaust, they have created through their chosen political impotence. Billy and the democratic crypts are about to be blown big time and it couldn't happen to more appropriate people!

Chapter 49
The Rad Race Car Drivers

I think people don't fully understand how wonderful race car drivers are, and what a spirited, playful, fun bunch they can be. Even though they take racing very, very seriously and know the importance of absolute attention to detail in the preparation of their cars, it still doesn't keep them from having fun in that work and in other things too. Around the Good race car drivers all sorts of funny, mysterious things happen. The bad people's cars can't start, or their windows won't open or sometimes maybe their trunk won't close. Occasionally, the bad guy's car is moved to a different spot so they can't find it in a parking lot, or their brand new Mercury ends up smelling like a skunk had spent the night there. Or if the bad guys are really, really bad, their cars may disappear altogether, to make it harder for evil people like Dan Havis to get to the meetings, where they plan to do the good race car drivers in. This playfulness on the part of the good guys, irritated the bejesus out of the bad guys, particularly the ones like Assr that have been made to walk and ride a bus to work, in punishment for trying to murder the nurse who told the world that Hord had done our patients in and now was trying to do us all in with contaminated water. Water that Hord's is still contaminating through dumping mercury and aluminum effluent into it daily. The fact that the good race car driver's were helping us, irritated the evil doers so much, that they murdered two of the Good Guys young sons at the same racetrack on the same turn, two months apart. The Concorde, New Hampshire track, where the Federal Health Care Financing Office was located to give HCFA and the race car drivers the message not to mess with Hords and the crypts big "bidness". But the Good Race Car drivers are resilient people, with many, many friends. The Truck people love them, as do the good Car people and others from all walks of life who know that a PHD or an MD or an MBA for that matter, does not necessarily a hero make. They also know that there are more felons that head corporations, than there are in Jackson prison. The only difference being, that the creeps from the crypts haven't been caught yet. But that work is in process too, the more the freaks react and retaliate against the Good Race car drivers, the clearer the trail becomes back to the keepers of the crypts. So Jac may change his name to "Jacques", but the connotation remains the same and everyone will still call him Jacass Assr, because that is what he is. I think I will have some pennants made up to fly from the aerial of my car, one with a donkey on it for Jacass, one with a carp on it for Troutman and one with a dead bird on it for the Hay Hords. Do you think when they see those flags flying, they'll get the picture, that everyone knows just how big of an ass they all are. It would appear at this point in time, that the only one who doesn't know it is themselves.

Just as Zero diddled while Detroit burned, the Dearborn Duds played with themselves while the Glass House shattered and they were left sitting on their ass in the grass, like Dorothy in Kansas wondering where their corporation went. The answer is that it went up in smoke my friends, you were so busy trying to kill off all your competition, to notice that no one will go near you now with a ten foot pole, because they don't want to be tainted with the stench of the dead people that follows you around like the corpses of the dead fish near Hord field. Every time you have an unexplained event happen, like legionella in your air system or broken water mains, or Lions and Tigers and Wings that can't win, no matter what, just think of all the people that you have screwed over in the past that are coming back to haunt you. Think of all the good people you have destroyed and why, and you will clearly understand that those unexplained events are fully explainable. It is the Curse of the Crypts coming back to vilify you and doing "bidness" in "Troit" will never be the same, again.

The world has got your number now, Triple 000 and they "Think Hord First Only on Autopsy!"

Just as Big Jimmy Hoffa sits between the goal posts deflecting Hords goals like flies, the spirits of the Good Race Car drivers sit in the passenger seat of each Hord car, with their foot on the brake and their hand on the throttle, just waiting to get even. The Terminator has joined them now, the result of another accident that was not an accident, a razored seat belt, just like they did to the safety harness of the World Health Care workers on the bridge. But Earnheart was a man with tremendous spirit, the spirit of a Manitou, so now the drivers for the keepers of the crypts truly won't stand a chance!

Chapter 50
The Duke of Pearl

I am trying to remember who became involved in what sequence. I met Pearl after the bad Drants, Cowman, Prasek, Querassi, Kerit and Sno finally revealed themselves, after they set up the Pink Bandit, sabotaged him so that their flunkie Artass Whortering could take his job. The Pink Bandit had done a wonderful job of making sure that we had all the documentation that we needed to fry the keepers of the Crypts, and as a result, the bad Drants were pooping their pants because they knew that if we went to court at this point they would be done like dinners. When Taresa, let slip the names of one of the bad Drants, I did a little ring work of my own, a little catering work. I prepared a lunch basket for one of the Drants, to let them know that I knew who they were, it would make it easier to ID them the next time around.

The second time I dropped off a little present, you would have roared, we had a repeat of the Orchestra hall keystone cops fire drill all over again, every security guard in the place came running and every Dr. including Pearl. Pearl was very easy to pick out in a crowd, first of all he was dressed in combat fatigues, not your usual hospital garb, and secondly he was a big guy. So a couple of weeks later, when I let slip on the phone that I had thrown the charts in the lake (which I hadn't), it was very easy to recognize who was standing on the lawn in front of my house, fishing, not for fish, but for charts. The whole lake was full of boats that day, fishing for charts. Pearl was an officer in the army, so we had an entire battalion out there with all of their little depth finders and infrared toys. Poor Pearl they had duped him too. The bad Drants hadn't told him the truth why they wanted the charts back, to cover up their complicity in the multiple deaths of the patients who had died from contaminated water at Hinai. They just told him, that it was important very, very important to recover those charts. So I let them sit out there all day in the hot sun, looking for the charts that lay scattered around my den and at the hospital where I worked. I figured we would let them make fools of themselves one more time, and let the whole world know how scared shitless they were of being discovered. Charles Wasson was in the army too and had fed Pearl some bullshit lines as well. Charles Wasson the freak who ran the city of Detroit's fire department, ran it into the ground that is, jeopardized the lives of his own men, so that he could scam the department's money for his own pocket, the ass! So after I found out who was screwing who, and that Pearl had protected some of the good doctors that the keepers of the crypts had tried to murder through accidents that were accidents, I called him up. He needed to understand the complexity of the dynamics that we were dealing with, so that he could protect himself and the good Doctors that were helping us and create balance for the bad army people that were in Wasson's group. Once he understood how dirty they played, he had special teams of residents protect the Docs who were patients in his hospital that the keepers of the crypts had tried to off. The doctors with subdural hematomas from falls that were not accidents and those who accidently had both their renal arteries severed after having minor surgery at one of the crypt's hospitals. Pearl saw that there was a war going on between good and evil and so he chose to protect the Good Guys, to keep the creeps from the crypts from finishing them off. Pearl is one of the Heroes too, an Oaktree man, who once he understood what was happening, did everything he could to protect his patients' because he does care, like Roger, he truly cares about the little people too!

Chapter 51
The Good Doctors' Revenge!

Just like the Good Race Car drivers have a tremendous sense of humour, so do some of the Good Doctors. Otherwise why do you think that so many of the evil-doers like Billy Hay Hord Jr. end up literally and figuratively speaking, not being able to wipe their own bums, after shoulder injuries? Or did you think it was just an unexplained event that so many of the animals who liked to watch pornography in Archie's bunker, are ending up reading Braille. Did you think that all the bad guys who are walking around with semi-colons and who have been carved a new assh..le, literally and figuratively speaking, were the result of an ass epidemic. Perhaps you thought the rash of impotence amongst those who thought themselves important, like Dick and Tom Swein was an act of divine intervention. Or that the piano players who can no longer pick up their pens to write, all suffered from the same malady. Did you think that all the Bad Guys suffering from the heart break of rosecea or psoriasis were just victims of a skin disease disorder. Or that the evil albinos who were all suffering from boils and impetigo, were victims of an odd epidemic? Or that 600 RCMP officers would get the poops simultaneously in Regina because of a fluke. Or that mad cow disease would come back to plague a few bad Windsor surgical Drants who thought they were sacred cows? Well then, you need to think again my friends, this is the Good Doctors' revenge for all of their friends who were injured and killed by the keepers of the crypts, and no one would do a f…king thing about it. The so called authorities were either paralysed into impotence or on the take, so the Good Doctors are taking things into their own hands, and there is not a thing that the creeps can do, because no one can prove anything and the crypts protect all malpractice. So in essence the Bad Guys are being screwed without a kiss by their own system that they've protected so long for other reasons. Talk about poetic justice, even Shakespeare couldn't have written a better outcome himself. And every time they hurt or kill one of the Good Guys, the retaliation is ten fold. You don't screw with the Good Guys who are trying to protect children and you don't murder their friends or their patients, or at least if you do, you don't get away with it for long!

There are many wonderful Docs, ERP's, Radiologists, Neurologists, Opthamoligists, Orthopods and Trauma Surgeons just to name a few, who risked their lives, are still risking their lives to protect the welfare and well being of those who can't protect themselves. The Good Docs who picked the original patient records out of the garbage for us, and who made our flyers to protect our patients and xeroxed hundreds and thousands of copies for us, xeroxed them all on PMC, Hord and Hinai's copiers with their own paper. These GOOD DOCS are heroes too, and create a tremendous amount of balance for the bad Drants who spend their time between surgical cases, playing with water piks and sitting with the piano players, altering medical records at 3 o'clock in the morning and injecting 72 year old beating victims with sedation to keep them unconscious, so they can't identify who beat them and why. YEA for the GOOD DOCS, they are Bravehearts, Warriors, Heroes and Oaktree Men and Women! YEA for them all! They make the evildoers and the keepers of the crypts look like piss ants that they truly are!

Chapter 52
The Unique U.N. and The Bosnians!

I knew very little about the U.N. and Bosnia as well, before this all started. But when we started seeing all the Bosnians who fled from their home land, ending up in hospital, dying from either cancer or kidney disease, it became once again very apparent that something was terribly, terribly wrong there. Tony Despeisner gave up the answer to that riddle too the US government under the direction of Clinton and Powell had authorized the use of shell casings that contained depleted uranium and plutonium, because it was cheaper. They knew fully that they were exposing their own troops, the troops of the rest of the United Nations Countries and the people of Bosnia; the women, the children, the babies to a deadly form of radiation that would kill them through myriad cancers and kidney disease. The bastards knew all along, but they didn't care. Clinton had never seen military service in his life and his Joint Chief of Staff was a political ant who led the military from his armchair. Sending other Generals into battle, Generals who would later succumb to the same brain cancers, lymphomas and leukemia's that were killing the men and women that they commanded. When the people of the US military started demanding answers, the official party line was don't ask, don't tell, we don't know, while all around them the US soldiers were dying like flies. The French noticed too and other countries who had UN troops stationed in Bosnia saw the same patterns; the clusters of cancers and nerve disease and other maladies of radiation poisoning, and when they demanded answers, the prick Clinton pulled the funding to the UN. He figured it would be harder for them to investigate him without money. But low and behold, another Hero surfaced there in the form of Ted Turner. He gave the UN the millions of dollars that Clinton had withheld, in one of his squeeze play attempts. Ted Turner gave the UN his own money, so that the UN could show the whole world what an ass Clinton truly was, he and his chair of the joint chiefs of staff, who knowingly and wilfully sentenced generations of Bosnians, US and UN troops to premature death, because they were too f..king cheap to supply them with munitions that weren't contaminated.

Clinton and his flunkie the chair of the joint chiefs of staff who thought nothing of sending their country to war, as a distraction, when Clinton had been found with his cigar up Lewinsky's crack, had done it one more time. Sending his own troops to war with radiation contaminated munitions merely to save a few bucks!

Assholes squared, is a very mild description for that pair. But what do you expect for someone who pardons felons as a ruse to expose 900 FBI officers. I'll tell you what to expect, absolutely nothing and then you'll never be disappointed. Clinton and his cabinet is the biggest political embarrassment the world has ever seen. And as the rest of the truth comes out, as to what was covered up under his and Powell's direction the story will only get worse. Powell who was so outraged that the U.S. had lost a seat on the U.N. Human Rights Commission. Well dah, what in the hell did they expect. A country that would poison 111 of it's own people with carbon monoxide on Lake Powell, simply so that a certain general could vacation in absolute silence, isn't much of a role model is it? Nor is one that contaminated Kuwait and their own troops with poison gas during Desert Storm. Powell needs to wake up and smell the coffee, the whole world knows about his façade of human rights and the perversity of his concept of diversity. It would appear now that the only one he is fooling is himself!

Chapter 53
The Gay Blades and the Terrific Transvestites

I think that gay people and he/she males are some of the most marvellously funny and courageous people I have ever met. No matter how grim the circumstances, they are still able to laugh and play and poke fun at the ludicrous behaviours of the people like the gruesome foursome, the ick club and the keepers of the crypts. When we were in Boston and the democrats were frostily ticked because they had lost the election, Richard was marvellous, teasing and playing and telling me about a play spoofing Maxie Pisser and the virgin Scary that was playing at the water front . When no one in the room would speak to us because we had given all the goods to the republicans to defeat the democratic crypts, Richard did, and so did the Ru Paul look alike that manned the front desk. The cab drivers were wonderful too, they knew all about the two tiered democratic system of health care, they were victims of it, spending half their income on health care insurance, so that when they or their loved ones became ill, they wouldn't go totally bankrupt.

I've often thought that transvestites would make the perfect secret service or FBI agents, they are so used to dressing up and putting make-up on that disguises would come like second nature to them. They wouldn't have to frost their hair or wear a toque. They could just pop on a wig and run, and since so many of them were used to living in the closet, keeping secrets would be no problem. They could hang out in any bathroom, male or female to hear the local scoop and talk to women about hair and harassment and men about business and baseball, and be very convincing at both. Best of all, they would have fun with it, you don't see too many gays who look so uptight that they would suck a brick out of a wall, if they leaned up against it. They could play little goblin tricks on the evil doers, things like stealing papers from their desks and taping secret meetings by masquerading as secretaries and bugging their undies literally and figuratively speaking by posing as check out counter staff. It would be so easy, they could be females, he-males or we males, whatever the occasion called for and they would be able to do it convincingly. I think Louis Freeh should give the thought serious consideration, and then maybe he would end up with fewer Buttercups who are on the take.

I know another gay couple who have the unique honour of doing Don Schmancy's hair and my hair too. Don's is a cheap date, the trick is to find it first, and after that cutting it is easy. Don tried to have them fry my hair, just like someone else did when I first came home, but they wouldn't do it. They wouldn't do it, because they knew him for who he was and what he was and so they blew him off like the dandruff that drifted from his eyebrows, the only visible hair on his head! It is so funny when the two of us are in the shop at the same time, the Don comes over and glares at me, paces the room and comes back and glares at me again. Then he mutters under his breath that Detroit had never heard of St. Louis or the Justice Department, until I came along, and all the time I sit there smiling sweetly and innocently at him. Trying so hard not to burst out laughing, if he only knew the extent of everything I did know, he would have a stroke for sure. I've told Randy the hairdresser to let me know when the Don comes in next so that I can bring him a copy of the book in which he and his crony Max Pisser are feature stars. But he won't get his copy until after the Justice Department reads it, which is right now, they are reading it as I type and I can't type as fast as them, but I can sure fill them in on the details, the linkages, like nobody's business. I can make them laugh a little too, just like I did three years ago before the keepers of the crypts and their allied frame came in with all their resources and shut us down. But this time George Bush and the new Justice Department are laughing all the way to the bank, because Bubba is gone, and there is a new federal Attorney General on the block, one that will get the job done right. Soon they will know about all of the double agents and about all the boys and girl's in Billy's cabinet who knew about them for a long time, but who didn't want to squeal, because they were guilty of the same damn things themselves! They were all selling U.S. secrets like half hour time slots at a sheets, no sheets motel! Happy Valentine's Day Fibbies, on our third anniversary, and like cherries on the vine this gets sweeter all the time!-with Luv Peaches (and here all this time, you thought that I was just another pretty bare bum!)

Chapter 54
The Zen of Zonjic and the Magical Musicians

If you should ask how does one keep their composure and their sanity in the face of such a horrendous mess, the answers there are relatively simple too. Never losing faith in the Good Guys is very important, as is the motivation that no one else should have to go through what we experienced, and finally music. Music that washes away the evilness of the world and let's us re-focus on the beautiful things in life. Whether it be rock and roll, or jazz, or Beethoven's 5th, music has the ability of transporting you from the very stinky realities that you have been caught up in, to a more peaceful plain that helps to remind you that the whole world isn't full of evil doers, whose only focus is doing in the other guy. That there are good people who create balance for that evilness, through their songs and compositions . Music is an eminently portable commodity. You can play it in the car when the freaks are chasing you down the freeway, or at home when the goons in the harbour are watching your every move, at work to humanize the sounds of the cacophony of alarms that exist in a hospital, in the boat, on the beach, anywhere, to put things back into perspective, a little reality tug of vibrating strings. Not the discordant notes of a bunch of creeps trying to do everyone in for a buck, but lyrical music, bouncy music, passionate music. Music that stirs your soul in a different way, a positive way, a light and enlightening way, so that no matter what the circumstances, you can put things in perspective and see things for what they really are. That there are truly very few discordant, out of tune strings, but when those evil people are in positions of power, the whole symphony is ruined, and what should have been a work of art, a tapestry, has become Grade 7 band practice, with all the petty intrigues, plots and sub-plots of a bunch of dummies, who have never emotionally made it beyond grade school. So in the face of the fools, I just turn the music up and tune them out, and let the world watch while the bad guys make asses of themselves, chasing the Good Guys, who were just trying to protect everyone, the bad included, from an early death caused by the contaminated cesspool of a water supply that the bad guys had created.

Music can be a very creative way for the Good Guys to communicate as well, when you know which songs are their favourites, it's easy to tell who is close by, just by the radio station being changed, and hearing the song that is played after the station change. Whenever you hear the theme song for the Titanic being played, you know that the Good Guys have whacked the bad guys one more time. Either through fining them big time for the contaminated water or suing them for wrongful roll-over deaths or through "older worker", class action lawsuits. When the evil doers murdered the agent that we called the "ICEMAN", they struck an ICEBURG, and now the keepers of the crypts are sinking millions and billions of dollars at a time. It gives new credence to the joke, " What do you call a thousand piano players at the bottom of the ocean?" Answer: "A good start!" The keepers of the crypts and their piano playing whores are riding the Titanic to the bottom of the ocean, while we sit back and clap. The bastards who allowed their own children to be murdered by the contaminated water, have been smoothly jazzed and they never even knew it! Assholes!

THE BAD GUYS REVISITED!

Chapter 55
The Revenge of Don Schmancy and Chippie McWhore

It seems that the Don didn't like my reference to the dandruff that drifts from his eyebrows and he didn't like me telling Chrysler's and GM that they were being stripped out financially by him, the Fendry Hord Mossad Company and the Ho's of Homerica. And Chippie didn't like me telling Roger, that Chippie had robbed him blind and destroyed the motor store. I wrecked some of their most lucrative schill games. So being the pricks squared that they are, they arranged for another accident that was not an accident. This time they tried to off my son-in-law David, by hiring some freak who almost ran my youngest daughter off the road three weeks before and did run a truck off the road, to try to do in David too. But they are so freaking stupid, they used the same car again, a 1988 red Honda Civic. They used a Civic, to make it look like it wasn't Hord related and then they gave it all away by having a fat fool threaten us in the tow truck parking lot. The fool was 6'4", 245 lbs, sandy brown hair, beer belly and was standing there with a stainless steel mug that had McWhore printed on the side. He wanted to make sure that I knew that his truck would be up my ass too, if we printed this story. I guess the Hords and Pissers and the Bunkers don't like people to know what murderers they are! Well guess what Buttercups, I will give you a real news flash, the whole world already knows what animals you are and you're just too damn stupid to realize it. Do you think I was the first one to think of the expression, "Think Hord First on Autopsy?" Your reputation is based upon 52 years of consistent practice; if you can't kill it, then cover it up! The only people you are fooling are yourselves. Do you think the workers at the Rouge plant, where 10 have been murdered and 22 injured in a year, don't know what your priorities are, when you allow them to die like flies because of your consistent negligence. While at the same time you plow the money that could have been used to create a safe workplace for them into building toilets for Hord field. I mean really give us a freaking break, do you really think any of us are even half as stupid as you are, assholes! So once again, instead of fixing the problems, you try to kill someone else off. A 25 year old father with a three year old daughter and a new baby on the way. You know that he has a graft on his aorta from a car accident that almost killed him when he was 12. That is why you picked him as a prime target, because he would be easy to off. You are so freaking low, that you would need a ladder to climb a postage stamp, and I would tell you to grab your balls and grow up and be men, but unfortunately you wouldn't be able to find them in your ruffled panties! Think Hord first my ass, only at the mortuary! Then to be even tackier you had Lincoln towing take the car to 1333 College Avenue, as a reminder that this is what happens to people who tell the truth, but guess what Chippie, they have some nice satellite views of both the accident and your fat freak threatening us. Just like they have of your creeps in the two white Hord Volvo trucks chasing me down the 401. Chippie, Chippie, will you and the Hords and Schmancy never learn, Billy Blow Job is gone and there is no one to protect your bullying tactics anymore. Do you think the Bushes want anything to do with your murders and your petty side shows, I don't think so. I don't think they appreciate the fact that the American and Canadian children here are being decimated by contaminated water, so that Barclay's bank, Fidelity Magellan, United Negro College fund, Max Pisser and his Mossad Muppets and a billionaire Egyptian, can accrue a few more pennies on their dividends this quarter. Tell Assr and Troutman to save the dramas for the rest of their queens and spend their time trying on their dainties instead, they are much more appropriate to their behaviour.

And a message to the Bunker boys, "Right now the law is clear to us, we must remain anonymous, my ass" . You are murdering your own children for Christ sake, WAKE UP! The only thing that being anonymous is going to get you and them is a berth in the Wayne County morgue!

June 2001 Postscript: Amazing, isn't it, the night before the Walkertown hearing in Windsor, David and my daughter Amy who is eight months pregnant are hit by a freak who ran a stop sign.! A Hord-related freak who ran a stop sign. Clearly they didn't want the Walkerton inquest to hear the Tony tapes that describe Hord, the Mossad muppet club and the Democrats' complicity in Walkerton, as well as many, many other atrocities! Too bad, how sad, but guess what freaks, a little road trip to Toronto will take care of that! Hitting a woman who is eight months pregnant, what a bunch of fucking freaks!

Chapter 56
Bribery Blue Damage Control!
(or the Fendry Hord Mossad Muppet Company!)

How do the Billy Hay Hords, Jackass and Troutman dig themselves out of the great big hole that they have dug themselves into! By throwing more money around of course. So women won't think that they could possibly drug and harass nurses, they throw ten million dollars at Smith College, and so Chrysler's won't look too closely at the fact that they stripped out the company financially with Homerica's and Pisser's help, they threw Chrysler's a little "bidness", by having them make diesel motors for the European Hexplorer market, so that poor Chrysler's can be blamed and share the legal liability too, when the damn things keep rolling over and killing people. To shut up the Boston democrats, so they don't look too closely into who murdered young John and Carolyn Kennedy and why, they throw fifty million dollars at Havad University, to buy them off like a bunch of cheap academic suits, to usage the embarrassment of their collusion with a bunch of freaks that are continuing to murder the 1,200,000 people here through environmental genocide. I have news for Hords; you are throwing the money at the wrong issues again assholes. Fix the freaking plants that are dumping mercury, aluminum and PCB's in the water supply and divorce yourself from Archie's bunker and BBB Benny and the Jets and then maybe, just maybe, you will start to get a modicum of credibility back. No one is impressed when you throw money around for all the wrong reasons, while at the same time you tell your employees to come to work at the Cleveland plant, but don't drink the water or wash your hands, because bummer, it might kill you! Why don't you spend some of the fifty million on fixing the water supply at the plant, but I guess that would never occur to you, because you don't give a flying f..k for your employees or anyone else for that matter, you prefer instead to bump them off or silence them like lambs, through your bullying and intimidation tactics. If you keep on acting like you are right now, the shards from all the damage that the glass house has caused, will continue to plague you, like shrapnel in your ass, because of you doing whatever, to whoever, whenever including Louis Freeh, and all the Good Guys and then lying about it with the Kurley gang. Do you really think anyone wants to buy a car from a mossad muppet company that murders children through environmentally caused cancers? I think you should start selling them with a bouquet of daffodils on the dashboard, so that people will have their own funeral flowers ready for their children, that you are killing! GM has gotten the hint and is cleaning up it's act, but you guys persist in pooping literally and figuratively speaking in the water supply, and then pretending that no one will notice the trail of brown stains, leads back to you. Guess what, even a blind man can smell it and Stevie Wonder could clearly see the decimation in the populations of thc citics of Windsor and Detroit. Read the obituaries, for Christ sake and the causal factors of all the deaths, they are like freaking phone books in length, with death after death, from environmentally caused cancers, birth defects and cardiac disease. Think Hord, Pisser and Archie first, only at autopsy and you will have truly made the, "Right Choice!" Mossad muppets and semi-colon puppets, responsible for the systematic environmental genocide of I,200,000 people, something I'm sure they're really proud of! Assholes!

Chapter 57
Requiem to Opus Deio

Have you ever noticed that the more fervent a person is in regard to their religion, that is usually because they are covering up some other issues big time!

I mean, look at Jim Baker, Jesse Jackass, Maxie Pisser and even Robert Hanssen, aka Ramone Garcia. All holier than thou, proselytizing bigots, who are dropping their drawers with one hand, while holding the bible or the torah in the other. And it certainly isn't just religion they are giving to their prospective converts, although some might say that it is holy rolling with a broader definition, that gives come on to me a new meaning. I was a little concerned when I read about the Opus Dei sect; it was almost like the rabid radical rabbi group with a twist, I mean these guys actually walk around with spiked bands on their thighs to inflict pain on themselves. I mean whoa, why not just go home and let your wife yell at you if you want to be abused or go to your corner dominatrix or better yet, go work for Hords, they would be happy to kick the crap out of you free of charge. Another dynamic that I don't understand, is that they are so fervently against abortion, that they murder the people who perform the procedures. Now isn't that an oxymoron, you kill someone because you are enraged that they are killing someone. Doesn't make too much sense does it. Nor does the fact that Cardinal Maida would blow sixty three million dollars on another mausoleum in Washington, while the children in his own backyard of Hamtramck are being decimated by the lead in their city's pipes. Why not give the city six million to fix the pipes and make the mausoleum just a tiny bit smaller. I'm sure Christ wouldn't care, he didn't even have indoor plumbing, so I'm sure a few less square feet aren't going to matter in his book. And please help me understand how the same churches that get their panties in a bunch over euthanasia, can so easily turn a blind eye to all the children who are dying from the local environmental holocaust. Why doesn't Cardinal Maida get pissed about that, murder is murder, is murder, what difference does it make, how it's done, unless maybe it's a matter of who pays for his golf games. And why doesn't the RCMP who jailed the farmer in Saskatchewan for killing his severely retarded daughter, do the same thing to the Ford executives who kill children with their defective cars, that they knew had serious safety problems for years. Why? What is the difference between Richard Gacey's serial killings or the killing of ten workers at the Rouge plant in the last year from negligence in safety issues? Dead is dead is dead, and murderers do wear three piece suits you know, some even wear police uniforms and FBI gear. Speaking of the FBI, since the world now knows that Ramone had a little tootsie of a stripper on the side, it makes you wonder how many of the Opus Dei boys are singing deio, under the sheets. I'm sure that he is not the only double agent who sings to support his extra curricular activities. Actually, I am positive that there are at least seven more that I know of that sing bass and soprano in more than one choir. I know because Tony Despeisner told me all about the children good and bad who play their freaky little games, hoping they won't get caught, or that Billy would bail them out. Apparently Ramone knew me too, or at least he thought he did and I knew his face. I recognized his picture right away, but couldn't for the life of me, understand why he had been following me around, unless maybe he wanted to convert me too, a scary thought! Then of course we have Mossad Max Pisser an icon of the Jewish community, who sits on the board of Homerica bank and plots with all the other flunkies, nine hundred and ninety-nine ways to keep polluting the environment and not get caught. The same Max who gives hundreds of millions of dollars to fight the, "holy war" in Israel, while he and his cronies literally and figuratively murder the children here through environmental genocide. Why not spend a few of those million on putting scrubbers on Detroit Edison's smoke stacks or putting holding tanks on Hords aluminum and mercury discharges? Hell no, his choice is to spend the money protecting a piece of dirt in the middle of nowhere. I think we should just sell him a chunk of land in Arizona, let him relocate a few holy rocks to it and call it a day. Maybe then the rest of the world would be spared the aggravation of their external fight over a piece of real estate that you could probably buy for a hundred bucks in the middle of the Nevada desert. But then again, that wouldn't allow Maxie's schill games to continue, of collecting money for the cause, his cause, and Don Schmancy's too!

I know lots of people from Detroit too; Catholics who would call you just to hear the sound of your voice, then masturbate while they had you on the phone. It was pretty easy to tell that they weren't making notes in their Franklin planner, when they called you from Gaylord or Traverse City or wherever, and made conversation in short little gasps and guttural groans. The same people who would follow you around at the office for a chance to look down your blouse or up your skirt. The same people who espoused their religion and wondered if all souls went to heaven in public, while they were doing all kinds of things in private to ensure them a road trip to hell. Boys from the accounts receivable club, who didn't like it when 5'6" brunettes, gave them back a little of their own. Who lost their sense of humour when the tales of the crypts went public and the holocaust of Hell Michigan hit the press. Funny how they lose their sense of humour so rapidly, once the public knows of their crimes. Their x-ray, Mickey Mouse and power pack jokes seem to lose their lustre as rapidly as Kurley, aka Archie is losing his hair!

Chapter 58
The End!

Now that we have cleaned Louis Freeh's contact lenses and cured him of his "tunnel" vision and blown the cigar smoke out of Buzzy Krongard's eyes(Buddy whose job it is to guard the kroner of mossad and the big investment banks!), I think we will ask God's help in explaining some U.S. miracles! Perhaps he can tell as how a wealthy man who used to sell soap and brushes, was able to get an "elective" cardiac transplant at the age of 71! I mean a lot more people are signing their donor cards, but it is pretty hard to donate your heart and still play the sax! He might also want to explain the one hundred and eleven injuries and deaths that have occurred on Lake Powell, every time a certain general vacations there! I know that some military men are filled with hot air, but carbon monoxide? That doesn't make sense! Or how about the $62,000,000 in nose candy (cocaine) that is missing from BBBBenny Napoleon, the "Troit" police chief's lock up? Sixty two million dollars worth is a lot of nose candy, did Benny give it out at Halloween, or maybe sprinkle it on the sidewalk in an altruistic move to save the city some money? Or did he sell it "hisself" to generate a little "rebenue" on the side, or maybe he blew it up his own nose, and that's the reason why the boys at "thirteen hunnit beaubein", don't know nuttin, now! Maybe that is the causal factor for his faulty memory about illegally locking up witnesses to murders that his Goons committed and his falsified crime statistics, missing evidence and faulty memory about all the bodies he has archived in the Packard plant! He don't know "nuttin" about "nuttin" any more, because his brain is so fried on high octane hash, that he can't even find his own ass to wipe it now! But God, (not the goons of Dennis), but the real God does know now, and with his new pair of Opus Deio, sexy specs, he will help us to solve the mysteries of Archie's bunker and Mossad Max Pisser's Temples of Doom! God might also be able to explain to us how Michael Powell can legally leak insider information to his friend Booby Pittman, so that both of them can rape Microsoft to AOL's benefit. But then again, maybe God doesn't deal in securities fraud and actually we have other priorities for him, like getting Cardinal

Maida off the golf course and behind the pulpit where he belongs. His job is to save souls literally and figuratively speaking and he needs to be doing his job and leading the action to stop the murdering of the children in this area from environmental genocide. Let him start by saving the 563 children in Windsor who have been afflicted with cancer, then move on to saving all the children who are still being pickled with formaldehyde at PMC's children's dialysis program! That will get him to heaven a hell of a lot quicker, than building another mausoleum in Washington. The world has too many crypts already, he should know that from living in Detroit. So why doesn't the church focus on leaving a real legacy, a living legacy. The church can think of it as revenue generation, if it saves the lives of the 30% of the children that are being decimated, that is 30% more parishioners that they will have in future. Perhaps if they think of it that way, they will grab their balls (other than the ones in their golf bag) and be true men of the cloth, instead of shepherds literally leading their flocks to slaughter. It would certainly enhance their image and their credibility!

Now that the tree has been shaken, the rotten fruit has fallen to the ground, littering Detroit like Dorothy's Kansas orchard after the tornado. The whole world knows that the once honourable Democratic Party of the United States of Mossad, Bejing, Moscow, is just a rouse. A political farce on a grand scale, a national scale, who sells out it's own people to corrupt big "bidness" for less than the price of a big mac! A party who lets it's children die like flies from environmental genocide in order to protect their own place at the political trough. Bigots who pretend to give benevolently to the causes of cancer, birth defects and heart disease, while guarding the data, like the crown jewels, that plainly shows they are the cause of the holocaust!

Our efforts haven't been in vain, the lid is off the crypt now, exposing all the corpses of those that have been poisoned by the mercury, aluminum, PCB's and PAH's from Zug Island and the Trenton River. And the footprints in the tomb clearly show who the trail of the carnage leads back to. Hords is in shambles, G.M. and Chrysler's have been rocked to their core and understand now too, that even the biggest bullies can easily become victims of their own recklessness, greed and quest for power. In just three short years, they have lost countless billions of dollars, their reputations and their pride. The corrupt kings of G.M. and Chrysler's are gone, and those at Hord are clinging to the sceptre for dear life, like leaches to the corpse of a whale after a typhoon. The whole world sees them clearly now, for who they are and what they are, monsters who murder children for a few cents a car and a few pennies a dividend.

The whole world sees, all because they murdered my friend. He didn't have a lot of money, but he had strength, courage and tremendous integrity. He came back six times to help us before they murdered him on Aug.13, 1999 and as a result, the U.S. economy has lost 666 billion dollars in the two years since his death. An eye for an eye, a tooth for a tooth, a billion dollars for his death and that of every child that they have murdered through the environmental genocide . Fitting retribution to the keepers of the crypts.

Now the tables have turned, the Good Guys are growing stronger everyday and blossoming, and the evildoers are crumbling in on themselves like the decayed flesh of all those they have murdered. My friend would be so proud, so proud of our efforts. He was my dearest, darling, handsome love; my Grandson will be named for him, Joel Matthew. He was the Agent they called the Iceman, whose heart melted for me. He was my history and physical doctor, to whom we told the history of the Crypts and that's why they killed him! But he didn't die in vain, just like Big Jimmy Hoffa sits between the goal posts on Hord field, my friend sits between the goal posts of the Red Wings net, and he too flicks away like flies, any goals the evil doers might aspire to. An eye for an eye, a tooth for a tooth, a child for a child and the life of a zero for each Hero that they murdered . A fitting creed for the third millennium and a fitting end to the Keepers of the Crypts!

THE END!